BRITISH LABOUR STRUGGLES: CONTEMPORARY PAMPHLETS 1727-1850

THE SPREAD OF MACHINERY

Five Pamphlets

1793-1806

Arno Press

A New York Times Company/New York 1972

Reprint Edition 1972 by Arno Press Inc.

Reprinted from copies in the Kress Library
Graduate School of Business Administration,
Harvard University

The imperfections found in this edition reflect defects in the originals which could not be eliminated.

BRITISH LABOUR STRUGGLES: CONTEMPORARY PAMPHLETS 1727-1850
ISBN for complete set: 0-405-04410-0

See last pages for complete listing.

Manufactured in the United States of America

Library of Congress Cataloging in Publication Data
Main entry under title:

The Spread of machinery.

(British labour struggles: contemporary pamphlets 1727-1850)
CONTENTS: A Letter to the landholders of the county of Wilts, on the alarming state of the poor [first published 1793].--Observations on the importance and necessity of introducing improved machinery into the woollen manufactory, by J. Anstie [first published 1803]. --Considerations upon a bill now before Parliament, for repealing (in substance) the whole code of laws respecting the woollen manufacture of Great Britain [first published 1803]. [etc.]
1. Wool trade and industry--Great Britain.
2. Machinery in industry. I. Series.
HD9901.5.S66 338.4'7'67730942 72-2545
ISBN 0-405-04437-2

Contents

A

LETTER

TO THE

LANDHOLDERS

OF THE

COUNTY OF WILTS,

ON THE

ALARMING STATE OF THE POOR.

SALISBURY:

PRINTED AND SOLD BY E. AND J. EASTON.

SOLD ALSO BY

E. HAROLD, MARLBOROUGH; AND T. SMITH, DEVIZES.

1793.

[*Price Threepence.*]

To the Landholders of the County of Wilts.

GENTLEMEN,

HOWEVER the public mind may be at this moment agitated by events of the greateſt general importance, yet it cannot be impertinent at any time, to call your thoughts to thoſe matters in which you are particularly concerned.

The Maintenance of the Poor is a ſubject which has, above all others, ever claimed your ſerious regard; and the expences attending it have been ſo heavy, and ſo ſeverely felt, that intereſt alone muſt have awakened your attention to it. But under the peculiar circumſtances of the preſent criſis, this matter becomes ſtill more urgent, and preſſes on your minds with double force. The introduction of the machines called, *Spinning Jennies*, into the woollen manufacture, has deprived the Poor of great part of their ſupport, and has thrown it entirely upon you. During the laſt winter you had ſome experience of the conſequences reſulting from this

 change;

change; the evil was then but partial, in the next it will be general. This is indeed the ſeaſon of induſtry and plenty; no one wants employment, nor gratuitous ſupport. But a wiſe government provides in peace againſt the exigencies of war; and if you have common prudence, Gentlemen, you will imitate it's policy, and not defer your attempts to counteract the impending miſchief, till it come upon you like an armed man. Your conduct, hitherto, has verified the old obſervation, that "Nobody attends to that, which is the buſineſs of all." I have ſeen you, with ſurpriſe, as inactive, and, apparently, indifferent, as if nothing had happened to the prejudice of your intereſts. It is true, you have complained, that poor-rates are very heavy; and to make them leſs ſo, you may have ſtinted the neceſſitous, and reduced their allowance. But the preſſure being removed, no marks of it remain. You ſeem to have forgotten, that the ſeaſon of diſtreſs will again return; and buſied in gathering your harveſt, you remember not how may mouths you muſt in future feed—A few weeks hence, I fear you will be rouſed from this inattention; and however you may now ſhut your eyes, and turn away your faces from the diſtant proſpects of calamity, yet, on a nearer view of it, you will behold a diſmal ſcene. For my own part, I can ſee nothing in the enſuing winter, but want and idleneſs, and their attendants, miſery and vice. The Poor will cry aloud for bread, and you will grant it with reluctance;

luctance; being without employment, they will have nothing to think of but their own neceſſities, and the means by which they may remove them: what thoſe means will be, I leave it to your judgment to ſurmiſe.— It is in truth a melancholy contemplation, but it is a neceſſary one; and my mind has ſo frequently recurred to it, that I cannot help endeavouring to impreſs the public with it's importance.

In conſidering this ſubject I have aſked myſelf the following queſtions:

Is it likely that the ſpinning machines will continue to ſupply the place of manual labour?

Or, Is it poſſible to prevent their continuing?

If it is not poſſible, what will be the conſequences?

Now it appears probable that the machinery will continue, unleſs government prevent it; or the intereſt of the manufacturers cauſe an alteration; or you, Gentlemen, apply yourſelves to that purpoſe with effect.

It is, or ought to be, the general tenor of government not to interfere with the jarring intereſts of trade, but to leave every individual to purſue his own advantages, with which the proſperity of the whole ſociety will uſually concur. To facilitate labour, and produce the greateſt poſſible quantity of any article, with the leaſt poſſible expence, is, moreover, the juſt object of it's encouragement. Were it therefore to interpoſe, with reſpect to the ſpinning machinery, it would contradict it's proper courſe of conduct. But ſtill more unlikely will be it's taking any part againſt the clothiers, in the preſent caſe, if it appear that it has in many inſtances actually departed from it's wiſe neutrality to promote their particular benefit. "Our woollen manufacturers," ſays an admirable writer, "have been more ſucceſsful than any other claſs of workmen, in perſuading the legiſlature, that the proſperity of the nation depended upon the ſucceſs and extenſion of their particular buſineſs. They have not only obtained a monopoly againſt the conſumers, by an abſolute prohibition of importing woollen cloths from any foreign country, but they have likewiſe obtained another monopoly againſt the ſheep-farmers and growers of wool, by a ſimilar prohibition of the exportation of live ſheep and wool*."—What favour then, nay, what juſtice can

* See Adam Smith's Wealth of Nations, vol. ii. p. 494—Where, enumerating the reſtrictions on the inland and coaſting trade, the cruel penalties incurred by the breach of them, and the vain pleas made uſe of to extort them from the legiſlature, he ſays, "Like the laws of Draco, theſe laws may be ſaid to be all written in blood." They have produced, however, the deſired effect, as in conſequence of them "the beſt Engliſh wool in England is generally below what wool of a very inferior quality commonly ſells for in the markets of Amſterdam."

you

you expect from the legislature, over whom the clothiers have gained so complete an ascendency? The abridgement of labour will be urged, besides, on all hands, as a beneficial scheme; and in many cases it undoubtedly is, where the labourers thrown out of one employment, can apply themselves to another. But where, as in the present instance, the work of twenty perhaps is performed by one, and the remaining nineteen are incapable of any other labour, and must be supported out of the general stock: I must confess, it surpasses my abilities to comprehend, how the public can derive any benefit from such a measure. But a discussion of this nature is foreign to my purpose, I only wish to shew, that it is very improbable government will contribute to the relief of the landed interest, by any interference in opposition to the mercantile.—What has been said, may be sufficient, on that head: nor will it be more difficult to shew, that the clothiers will not, of choice, restore the spinning work to those, whom they have deprived of it.

However the public interest may be affected by the machinery, their private interest is undoubtedly advanced. The former expence of spinning a pound of wool varied from twelve to fourteenpence, though it has been considerably more; it is said, that it is now spun for sevenpence. Here then the advantage is evident: But probably the manufacturers lowered the price of cloth in proportion to this saving, and the more ſo, as their plea for introducing machines was, that they were underſold by the clothiers in Yorkſhire. *— Did they ſo ?—The drapers will give you a different account; for at the very time the expence of manufacturing was ſo much leſſened, they availed themſelves of a momentary riſe in the price of wool, and advanced ſuperfine cloth, a ſhilling a yard: Nor have they again reduced it ſince the value of wool has fallen to it's preſent price.—But how could they do this?—Why?—becauſe the importation of foreign cloth is prohibited, and, owing to a concurrence of circumſtances, the exportation was very great: in ſhort, they had the public in their power. Never, perhaps, were the clothiers' profits ſo large, as when they could not afford to give the cuſtomary wages †.—But ſome of them have gene-

* The futility of this plea will appear likewiſe from the conduct of two or three liberal manufacturers, who have never adopted the machinery, and continue to employ the Poor—Such men deſerve the public thanks—they are the friends of the community, and may juſtly claim it's approbation.

† "In reality, high profits tend much more to raiſe the price of work, than high wages;" ſays the before-mentioned author; of this he gives a very full and ſatisfactory explanation, but of too great length to be inſerted here; and concludes thus: "In raiſing the price of commodities, the riſe of wages operates in the ſame manner as ſimple intereſt does in the accumulation of debt—The riſe of profit operates like compound intereſt. Our merchants, and maſter manufacturers, complain much of the bad effects of high wages, in raiſing the price, and thereby leſſening the ſale of their goods both at home and abroad. They ſay nothing concerning

generoufly declined the ufe of the Spinning Jennies, upon condition, that the farmers will give them ten per cent. for allowing the Poor a fufficient quantity of fpinning work at a very low rate. Others, whofe capitals are weak, have had affurance enough to afk a loan without intereft for granting the fame favour.

So, Gentlemen, you are content to pay the workmen's wages, and afterwards to buy their manufacture at an advanced price! You give the clothiers a premium for allowing you to fpin the wool that grows upon your own eftates!—They have all the gain, and you have all the lofs!—You can hope for relief, therefore, only from your own exertions.

It feems, on the firft view of the bufinefs, that fomething may be done; if the evil cannot be removed, it may certainly be leffened. But what is the moft proper way to fet about it!—I feel myfelf incompetent to decide! I have neither the practical information of a man of bufinefs, nor the fpeculative knowledge of a theorift: I merely wifh to excite attention to this mo-

cerning the bad effects of high profits—they are filent with regard to the pernicious effects of their own gains—they complain only of thofe of other people."

Wealth of Nations, vol. i. p. 149.

mentous

mentous object.—I find the idea of a combination of parishes, who shall establish a manufactory, and by that means employ the spinners, is thought liable to many objections: The principal of which appears to be, that such a manufactory would multiply the poor, and increase the rates wherever it was undertaken *.— But I think the principal objection is, what will hold equally against any other scheme, viz. The want of energy and union among the landholders, and their inaptitude for any extensive plan of general œconomy, and their dread of spending one farthing more than will conduce to their own particular advantage. If these causes do not prevent every salutary measure, there is one method which seems practicable: It is, the spinning of wool only, and then selling it in yarn; in short, the business of a market-spinner. Were a certain number of neighbouring parishes to unite, and contribute to the support of a person conversant in the business of sorting and combing wool, &c. and to supply him with a sufficient quantity of it to employ the Poor who are destitute of work,—this would at once preclude the difficulties which might attend the higher stages of the manufacture, and afford them all the advantages they could derive from it. A large stock would certainly be necessary; but that would be only

* The parish of Kennet, singly, has established a manufactory, and found it attended with tolerable success.

paying the poor-rates for one year in advance; if no profits accrued from it, yet the immenſe expence of ſupporting an unemployed Poor, would be avoided. There is no doubt, but there would be a ready market for the yarn, if well ſpun, which might be eaſily ſecured.—This, however, is only a looſe hint, which others may improve on: nor do I preſume to fix on the moſt practicable remedy for the evil I complain of. I would rather adviſe you, Gentlemen, to petition the High-Sheriff to call a general meeting, where the chief perſons in the county may aſſemble, and deſire ſome method of obtaining relief. The men of property that enrich, and of talents that adorn, this county, will, if they exert themſelves, certainly produce ſome good effect. And are they not called upon by every motive which ought to influence them in their individual and ſocial character? as they conſult their own intereſts, the cauſe of humanity, and the welfare of their country. For unleſs ſome active ſteps are taken to reſtore employment to the Poor, the conſequences muſt be dreadful; whether conſidered in a political or moral view.

That which will naturally ſtrike you firſt, is the immenſe burthen which will be added to the poor-rates, already ſo heavy, as ſcarely to be borne—An induſtrious mother of a family would, when ſpinning work was plenty, earn, with the aſſiſtance of her children, four or five ſhillings a week, which, added to the huſband's

wages,

wages, provided a tolerable maintenance; ſingle women were able to ſupport themſelves: but neither women nor children can now derive any ſubſiſtance from their own labour, except during the ſummer.—Now only calculate how much you muſt contribute to their ſupport; and then judge, whether you are not deeply intereſted in this affair. I know a pariſh which paid eight pounds a week, laſt winter, to the Poor who were deprived of employment: the Poor, I need not ſay, were greatly diſtreſſed with this allowance—Overſeers have complained that their rates were doubled; and yet, during that time, ſome ſpinning work was ſtill to be procured. Theſe circumſtances will give a faint idea of the miſchiefs which muſt ariſe, when it is entirely ſuppreſſed. The landholders will be exceedingly burthened; and yet the ſufferings of the indigent will not be relieved. The diſtreſſes which they have already undergone, notwithſtanding the extraordinary aſſiſtance which have been granted to them, will furniſh a benevolent mind with the ſtrongeſt arguments for an effective interference in their favour—The proſpect before them is ſtill more threatening: ſome relief they muſt, and will, obtain; but not enough.—We all know how unwilling every man is to part with the fruits of induſtry and toil; and certainly the yeomanry form no exception to this general obſervation. The juſtices may be applied to, but will adminiſter relief with a ſparing hand—they are all men of landed property, and will be fearful of leſſening it's it's value. But ſhould they be ever ſo diſintereſted, they muſt not ruin the farmer, by giving to the pauper—they cannot ſatisfy both; and, probably, will ſatisfy neither. It may be thought, that a farmer may riſe the price of his commodity, in proportion to his expence; but this is impoſſible:—" When a trader finds his profits inſufficient, he withdraws a part of his ſtock, and ſupplies the market more ſparingly, and the price immediately riſes. But the farmer muſt employ a ſufficient ſtock to cultivate the land for which he pays rent. It can never be his intereſt to diminiſh the quantity of his produce, nor, conſequently, to ſupply the market more ſparingly than before: He will never be able, therefore, to raiſe the price of his produce ſo as to reimburſe himſelf, by throwing the final payment on the conſumer *."

You ſee, Gentlemen, the evils that threaten you, and the hardſhips that will afflict the Poor—Want and miſery will be their portion—vice and immorality their inheritance. They will not be ſo much unemployed, as miſemployed. Idleneſs and ignorance were never yet united, without producing their natural fruits—riot, theft, and every iniquity. The moſt profligate will clamour, and complain the moſt.—I cannot but foreſee more fatal conſequences likely to enſue from this event, than from all the ſeditious writings and attempts

* Wealth of Nations, vol. iii, p. 307.

which

which have excited the public indignation. The cries of hunger will be heard—If entreaties are of no avail, they will be turned into demands; and demands and violence are uſually not far diſtant. What apt inſtruments will a ſtarving and outrageous rabble be for the enemies of our conſtitution to employ! How will they miſconſtrue the cauſes that gave birth to theſe diſtreſſes! How will they fan the flame of inſurrection, and direct it's fury!—Was it not lately aſſerted by a leading Member of the Houſe of Commons, "That three thouſand people in Wiltſhire were deprived of employment by the war?" Had he ſaid, one hundred thouſand were deprived of it by the ſpinning machines, he would have been nearer the truth. I leave the original author of this miſrepreſentation, to his own reflections; but thank him for having ſhewn, how eaſily ſuch circumſtances may be diſtorted to favour ſiniſter purpoſes.

Can you conſider all theſe matters, Gentlemen, with coolneſs and indifference? Can you look forward to the dimunition of your Property, to the evils that muſt afflict the Poor, and the dangers that may befal your Country—without feeling ſome concern—without making a ſingle effort to avert this complicated miſchief?

I am, Gentlemen, &c. &c.

POSTSCRIPT.

THIS Letter was intended to be ſent the Printer of the Saliſbury Paper, but appearing too long for inſertion, is preſented to the Public in it's preſent form. It is addreſſed to the LAND-HOLDERS *only, as to thoſe who have been eye-witneſſes of the diſtreſſes of the Poor, and of the cauſe which has occaſioned them; and having felt the burthens conſequently impoſed, muſt be ſenſible how much they are concerned to ſhake them off. To thoſe owners of lands, who are not occupiers, no diſtinct application is made, becauſe the intereſts of both are ſo intimately and inſeparably united. No one can be ſo ſhort-ſighted, as not to look beyond a ten-years leaſe; or ſo thoughtleſs as not to reflect, that when it expires, if things continue as they are, his eſtate will return upon his hands loaded*

loaded with additional incumbrances. "The farmer must have his reasonable profit, otherwise he must give up the trade. After the imposition of a tax of this kind, he can get this reasonable profit only by paying less rent to the landlord. The more he is obliged to pay in the way of tax, the less he can afford to pay in the way of rent. A tax of this kind imposed during the currency of a lease, may, no doubt, distress or ruin the farmer; upon the renewal of the lease, it must always fall upon the landlord."*

* Wealth of Nations, vol. iii. p. 308—The author is reasoning against a tax on the profits of Farming stock—the argument holds equally against Poor-rates, which are virtually a tax of the same nature.

FINIS.

OBSERVATIONS

ON THE

IMPORTANCE AND NECESSITY,

OF INTRODUCING

IMPROVED MACHINERY

INTO THE

WOOLLEN MANUFACTORY.

Price Two Shillings and Six-pence.

OBSERVATIONS

ON THE

IMPORTANCE AND NECESSITY OF INTRODUCING

IMPROVED MACHINERY

INTO THE

Woollen Manufactory;

More particularly as it respects the Interests

OF THE

COUNTIES OF WILTS, GLOUCESTER, AND SOMERSET;

WITH

General Remarks on the present Application to Parliament,

BY THE MANUFACTURERS,

For the Repeal of several of the existing Laws.

IN

A LETTER,

Addressed to

The Right Honourable Lord Henry Pettey.

BY JOHN ANSTIE,

Chairman to the General Wool Meeting in the Year 1788.

LONDON:

Printed by C. Stower, Charles Street, Hatton Garden, for the Author,

AND

SOLD BY STOCKDALE, PICCADILLY; VIDLER, 349, STRAND; [illegible]RINGTON, WHITEHALL; AND RICHARDSON, ROYAL-EXCHANGE.

1803.

My Lord,

THE original deſign of committing to writing Obſervations on the preſent application to Parliament, by the Clothiers, for repealing moſt of the preſent laws for regulating the Woollen Buſineſs, was, in the intention of the writer, confined within a narrow compaſs; and was intended merely for the private inſpection of the Marquis of Lanſdown, previous to the ſubjects coming under the diſcuſſion of the Houſe of Lords.

Though no one can be more ſenſible than the perſon who now addreſſes your Lordſhip, of the extent of knowledge, reſpecting trade and commerce, poſſeſſed by Lord Lansdown, yet he was fully convinced, remarks on a ſubject of importance to the Cloathing Busineſs by a perſon honored originally with his Lordſhips voluntary notice, as a *Woollen* Manufacturer, and continued, without interruption, for near twenty years paſt, would not be deemed unworthy his Lordſhips attention. In purſuing his deſign, the writer found his ideas expand much more than he expected, and he was inſenſibly led to enlarge, till he found it neceſſary to abandon his firſt intention.

It appearing poſſible, that he might be able to produce ſomething like a complete view of the ſubject, he continued, as opportunity offered, to commit his ideas to writing.

In his progreſs, finding that to be *probable* which at firſt he deemed merely *possible*, he was, by a particular circumstance that came to his knowledge, led to ſuppoſe that the publiſhing his remarks might anſwer ſome valuable purpoſes.

By elucidating the ſubject, not by theoretical reaſoning only, but principally by concluſions drawn from abſolute facts, he ſuppoſed it might be in his power to obviate the effects of partial information, which he knew had been communicated to reſpectable Members of the Houſe of Commons—and in ſome degree at leaſt be inſtrumental in removing the prejudices of thoſe perſons, who are adverſe to the further introduction of machinery into the woollen buſineſs.

He alſo entertained the hope that it might be poſſible to adapt his arguments in ſuch a manner to the different points he ſhould conſider in diſcuſſing the ſubject, as muſt have no inconſiderable effect in calming the ferment, which at preſent ſo violently agitates the minds of the work people in the county of Wilts, by ſhewing how much *their own real future interest* is connected with the application to the Legiſlature.

Encouraged by the *friend of his youth**, the writer now feels himſelf willing to ſubmit his free and unbiaſſed thoughts on the ſubject that has engaged his attention, to the candid inſpection of all thoſe perſons who may be diſpoſed to intereſt themſelves in its conſideration.

The more the ſubject is attended to in all its conſequences, the more he believes it will be apparent that it is not merely of a *local nature,* but that the eſſential intereſts of the *trade of the kingdom* are materially concerned in it.

Were the particular reaſons aſſigned that has impelled the writer to addreſs your Lordſhip on this occaſion, he might incur the imputation of being governed by a motive very different from what he really feels; he therefore contents himſelf with the pro-

* Thomas Fassett, Esq. of Surborton.

priety of conveying his thoughts to the public, on a ſubject that has repeatedly engaged his attention—under the ſanction of your Lordſhip, as a *Member of the Select Committee*, appointed by the Houſe of Commons, and as a Nobleman *particularly interested* in the welfare of the *county of Wilts*.

In the year 1791, ſerious apprehenſions were entertained of the ſhearmen, and other perſons employed in the woollen manufactory, committing freſh outrages on the property of thoſe Clothiers, in the county of Wilts, who at that time had more generally begun to ſcribble wool by machinery.

Obſervations were at that time drawn up, to recommend the adoption of precautionary meaſures, not only for the protection of the Clothiers, but alſo for ſatisfying the minds of the work people, by ſhewing the *absolute necessity* of an extenſion of the uſe of Machinery in the woollen buſineſs, to preſerve that *trade in the county of Wilts.* These observations, sanctioned by the approbation of a respectable clergyman, one of the Magistrates of the county, were delivered by him to the Justices of the Peace, when assembled at the county sessions.

Recent events appear to demonstrate that an application of the regulations, adopted by the Magistrates of the county of Lancaster, when machinery began to be more generally introduced into the cotton manufactories, and which were particularly noticed in the Observations, submitted to the consideration of the Magistrates of the county of Wilts, would have been equally beneficial to the peace of the county of Wilts.

The contest between the Clothiers of the counties of Wilts, Gloucester, and Somerset, and the work people, is now brought to a crisis, by the application of the former to the Legislature for a repeal of the obsolete laws, respecting the woollen manufacture, particularly those on which the latter ground their opposition to the further progress of machinery, in diminishing labour.

That the woollen business in the county of Wilts, in particular, is at present in a depressed state cannot possibly be denied; and no one, who feels himself in the least interested in its prosperity, can be insensible to the event of the present disputes between the Clothiers and the work people.

No doubt there are a considerable number of persons of respectability in the three counties, who, though they must condemn the flagitious conduct of some individuals among the work people, and who must also lament the destruction of property, which eventually must be borne by the *Hundreds* in which the factories are situated, yet still continue to consider the introduction of machinery into the woollen trade as unfriendly to the general intereſt, and peculiarly injurious to the poor.—It muſt therefore be an object of conſequence to point out, in a clear and ſatisfactory manner, the real neceſſity of the interference of the Legiſlature at the preſent moment, to enable the Clothiers to bring forward, without riſk, thoſe improvements in machinery, which exiſting circumſtances imperiouſly demand.

If ſuch perſons who are at preſent inimical to the plan of the Clothiers, can be led to ſee its utility, not only in promoting the general *interest of the nation,* but alſo in ſecuring *particular benefits to the manufactoring counties*—their active influence in reconciling the work people to what is propoſed to be obtained, must operate in the most beneficial manner, to prevent any future hoſtile oppoſition.

The following obſervations will, it is hoped, in ſome meaſure, facilitate ſuch a happy event—or at leaſt they will prepare the way for a more *accurate* and *satisfactory* inveſtigation of the ſubject.

It is to be lamented that general improvements are too often attended with partial evils, if very peculiarly fortunate circumſtances do not occur to prevent their operation. Precautionary meaſures to guard againſt ſuch evils, by thoſe perſons who poſſeſs the power, as well as feel the inclination, to mitigate them, muſt therefore, at all times, be highly expedient. The reſolutions of the county of Lancaſter, juſt referred to, may juſtly be conſidered as worthy of imitation, in all caſes of a ſimilar nature—and the utility to the public at large, of giving full ſcope to the improvements of the human intellect, in deviſing means for extending the uſe of machinery, muſt be evident, from the rapid increaſe of the cotton trade, to which thoſe reſolutions referred.

The introduction of almost all kinds of machines into the woollen manufactory in the county of York, without obstruction, from the work people, originated in very favourable circumstances, which, if necessary, could easily be pointed out; and the rapid increase of the manufacture, since the more general improvements of machinery in that county, must evince the superior benefits derived by those persons, who are enabled to reap the advantages of ingenious discoveries for diminishing labour.

The little opposition made in the county of Gloucester, by the work people, to the improvements in the manufacture, till the shearing frames were attempted to be introduced, must also be attributed principally to favourable existing causes, though the increase of the woollen trade of that county, must perhaps be considered as originating principally in the *superior mode* of *dressing* ſuperfine cloths*.

Unfortunately, ſo far from there being any thing exiſting in the ſtate of the county of Wilts, particularly favourable to the introduction of machinery, when the improved mode of ſpinning began to prevail, that circumſtances of a local nature were peculiarly adverſe to its being introduced into that county.

A large diſtrict of the county, where no manufactories were eſtabliſhed, depended almoſt entirely on ſpinning of wool, carried there not only by the clothiers from Bradford, Trowbridge, Devizes, Melkſham, &c. but even from ſome of the manufacturers in the

* In a subordinate degree, the alteration that took place some years back, in gradually making medley cloths in that county, no doubt contributed to increase the manufacture—but without the aid of superior dressing, comparatively little progress would have been made in manufacturing cloths, dyed in wool, (called medleys), though prosecuted with spirit by individual manufacturers.

counties of Gloucefter, and Somerfet.— When, therefore, the far greater part of the Clothiers ceafed from carrying wool for fpinning to that diftrict. The fpinners were thrown out of employ, and no previous meafure having been taken, no adequate fubftitute could be found for the lofs fuftained by them.

Their fituation at that time was truly diftreffing to the feelings of humanity—and the warmeft advocate for the introduction of machinery, if not abfolutely deftitute of all kind of commiferation for the fufferings of others, could not but be anxioufly defirous of devifing means for their being employed*.

* In conversation with the late Sir Richard Arkwright, many years past, the writer was led to enquire of him, what immediate effects followed from the extension of spinning cotton by the aid of machines by water, and in what manner the people were kept employed. He informed him the spinners in that part of the county, when the spinning was principally *done, by proper regulations having been adopted, were almost immediately* engaged, either in weaving or some other branches of the

To excite impartial attention to the arguments which will be adduced, to ſhew the abſolute neceſſity of progreſſive improvements in machinery, under *present existing circumstances*, it cannot be deemed improper to conſider more fully, the conſequences that muſt have followed, by checking the ſpirit of improvement, in the county of Wilts, in particular, even under all the diſadvantages it was expoſed to, from the local circumſtances of the ſpinning diſtrict, ſo very different in this reſpect from either the counties of Glouceſter or Somerſet.

If theſe circumſtances can be ſatisfactorily pointed out, it will contribute to prepare the mind to proceed with ſome degree of ſatisfaction, to the more enlarged conſiderationof the general ſubject,

business. The adoption of *similar measures* in the county of Wilts, the writer hoped would have produced *similar effects.* Other adverse circumstances, however, occurred some time after the general introduction of machinery, unfavourable to the exertions of public spirit.

Permit the writer, my Lord, therefore, to ſolicit your Lordſhip's particular attention, as well as that of the reader's in general, to what will follow. As he is himſelf fully convinced, the *grand source* from whence has ſprung the opinions generally entertained in the county of Wilts, adverſe to the utility of machinery, muſt be traced to its particular and diſcriminating features, as a manufactur-ing county.

Had there been no poſſibility of finding other modes of employing the people in the ſpinning diſtrict, even for a very conſiderable ſpace of time, when they were deprived of their uſual means of ſupport, the expence of their maintenance by parochial aid, muſt in-deed have been a melancholy conſideration to every perſon poſſeſſing liberal and generous ſentiments; ſtill, however, the aid would have been *comparatively* ſmall, when con-traſted with the almoſt *total loss of the*

Woollen Trade, throughout the whole county*.

This, in all human probability, muſt have been the neceſſary conſequence *eventually,* had not ſpinning by machinery, on the *improved* machines (as in fact the ſpinning turn, ſtrictly ſpeaking, must be conſidered as a machine), as well as ſcribbling and carding wool, by different and more expeditious modes been introduced into that county.

In vain will a *town,* a *country,* or a *nation,* hope to preſerve its manufactures, if a lethargic torpor binds it to an adherence to old modes, when *other towns, counties,* or *nations,* animated with a ſpirit of improvement, have found out new methods of manufacturing goods of a ſimilar quality, on cheaper principles.

* On a fine day in summer, to view the aged matron carding the wool, before the door of the rural cottage; the young children either handing the fleecy rolls to their parents,

Very peculiar and favourable circumſtances in the manners of the people and in the climate, may indeed, for a long period of time, enable one part of the globe to adhere to its ancient and prevalent principles of manufacturing particular articles, without experiencing any conſiderable detriment. Such, in a peculiar degree, is the caſe of the natives of the Eaſt Indies.

Local advantages, enjoyed by one country in Europe over another, may alſo procure a ſuperiority in ſome particular article. Such, for inſtance, is the manufacture of *Cambries* in France, and in the Netherlands.

or equally employed in drawing out the lengthening thread, while the chearful song of the healthy village damsels, uniting with the whirring spindle, completed the rural harmony, was undoubtedly a scene, with which the *feelings of humanity*, must have been peculiarly gratified. Truly happy must that person, or those persons be, who can be instrumental in restoring employment, though in a different manner, to every inhabitant of a cottage, in that district!

It muſt however be admitted, as a general fundamental principle, that even under ſome particular diſadvantages, that nation which has taken the lead in improvements in manufacture, if not checked by improper reſtrictions of the Legiſlature, ſuggeſted either by intereſted individuals, or extorted by the clamours of the work people, will proceed, in a regular ſtate of progreſſion, *in maintaining* its ſuperiority.

Let this mode of reaſoning be applied to the comparative ſtate of the county of Wilts with the county of York, when the application of jennies for ſpinning, and afterwards, when ſcribbling and carding machines in the Clothing Trade began to be apparently beneficial.

The price of proviſions was, at that time, more in favour of the county of York, than probably it is now, and conſequently labour alſo might be comparatively cheaper than at present. Still, however, the county of Wilts, from having been much longer the seat of the finer woollen manufactures, possessed many superior advantages—and, most certainly, the prevalent opinion of the purchasers of fine goods was, *at that time,* particularly favourable to the manufacturers in the county of Wilts.

Had a *liberal and generous spirit of improvement,* combined with a *just regard to the interest of the common people,* immediately taken place, the county of Wilts would not, at this time, have been in danger of losing that superiority in the finer manufactures, particularly in the manufacture of medley cloths, she has so long enjoyed.

Such a favourable state of things must have prevented the distress of the poor (at least as unconnected with unfavourable circumstances, from peculiar causes;) the destruction of property, both in the counties of

Wilts and Somerſet, could not have happened; *infatuated individuals,* by becoming the victims of miſguided zeal, would not have been expoſed to the juſt cognizance of the laws of their country; the *morals of the work people* not injured by vicious examples, might have remained uncontaminated by the perpetration of acts of violence, and ſome of the *principal manufacturers,* would not have been, as they ſeriouſly are at this moment, diſpoſed to forſake the two counties, if effectual meaſures are not adopted by the Legiſlature for their protection. Unfavourable as is the preſent ſtate of the county of Wilts in particular, in point of the woollen manufacture, how much more melancholy would it have been, had *no improvements* taken place, either in ſpinning, carding, or ſcribbling?

Perſons of humanity, feeling for the diſtreſſes of the poor, but by no means conſidering the firſt principles neceſſary to promote manufactures, and miſtaking the proper application of legiſlative acts reſpecting trade, might have applied to the wiſdom of the Legiſlature, to ſuppreſs the uſe of the new machinery *entirely*, not only in the county of Wilts, but throughout the whole kingdom.

This was propoſed by perſons of reſpectability; and had the opinion of the county been taken at that time, there can ſcarcely be a doubt but that it would have been in favour of the meaſure*.

Such an application to Parliament would have alarmed the ſmall manufacturers of Yorkſhire (to ſay nothing of Glouceſterſhire alſo); enlightened by experience, as to the

* The concurrence of unfavourable circumstances, which it might be difficult, even with the utmost caution, to glance at, did, however, most assuredly operate in influencing the judgments of persons of real understanding, and of different political principles, to attribute merely *local* or *accidental effects*, as they respected the miseries of the poor, to the introduction of *machinery simply*.

abſolute utility of their carding machines and jennies for ſpinning, and joined by the more opulent manufacturers and merchants, in that county, as well as in other parts of the kingdom, the table of the Houſe of Commons would have been loaded with petitions, in oppoſition to ſuch an application.

That in ſome counties, excluſive of Wiltſhire, Somerſetſhire, and Glouceſter, a petition to this effect would have been approved of by particular towns, will not be queſtioned*. But certainly it cannot be difficult to determine what the deciſion of the Houſe of Commons would have been on *such an occasion,* when the ſubject came under diſcuſſion.

Suppoſing, however, the prejudices of the work people, ſanctioned in general by the

* The writer is not, at present, prepared to speak to the actual state of the county of Devon; but if not greatly mistaken, no part of the kingdom would be more benefited by the introduction of machinery.

opinion of a great majority of landholders and other perſons, in the county of Wilts, at the time we are now contemplating, ſhould have ſo far intimidated the manufacturers, as to prevent even the *introduction of spinning*, by machines, till the determination of the Legiſlature could be known, what muſt have been the conſequence?

The Clothiers, fully convinced of the impoſſibility of meeting their *rivals* at market, without incurring conſiderable loſs, would, during the application to Parliament, manufacture as few goods as poſſible; and even ſuppoſing their patience should have laſted till the determination of the Legiſlature could be known, ſuch would have been the miſerable ſtate of the county, from the conflict of contending paſſions, and ſo difficult the reſumption of machinery, on an extended ſcale, that many opulent manufacturers might have been diſpoſed to ſeek more favourable ſituations, and even thoſe that remained would have had to contend with all the disadvantages attendant on a *dispirited* and *dissatisfied* class of work people*.

But, to proceed to more extensive views of the subject. Supposing the county of Wilts to have been actually reduced to this melancholy situation, respecting the woollen manufacture, yet still she might, in time, have experienced a recovery, by the gradual removal of this, as well as by the substitution of some other manufactures—and even at the very worst, the *nation*—considered in relation to all its parts, could not be said to be injured, as what the county of Wilts, by its improper conduct might have lost in the woollen manufacture, would have been gained by the county of York, as well as by other counties.

* Should any Yorkshire Manufacturer be disposed to peruse these Observations, let him suspend any unfavourable opinion arising in his mind from the mere use of the term *Rivals*, till the conclusion.

Let it however be confidered what the actual ftate *of the whole nation* would be, fhould fuch principles prevail as would tend to check and debilitate the general fpirit of improvement in the woollen manufactures?

To anfwer this queftion, in a manner fatisfactory to perfons of an enlightened underftanding, *the relative situation* of the different nations of Europe, compared with the prefent ftate of the united kingdoms of England and Ireland muft be confidered.

The immenfe expence of a long and fevere conteft with our enemies, has unavoidably burthened the nation with very heavy taxes, and this has neceffarily enhanced the price of provifions, confequently the price of labour has been materially affected.

It is therefore abfolutely impoffible our manufactures can be fent to foreign markets, on fuch terms as will enable this nation to maintain the fuperiority it at prefent poffeffes over our rivals, with whom the price of labour is much cheaper, by any other modes than by thofe which have hitherto effected it*.

Though it would be eafy to point out the beneficial confequences neceffarily following the application of machinery in other branches thefe Obfervations will be confined more immediately to the Woollen Manufacture†. At

* The perfect security of personal property arising from the *very nature* of our admirable form of Government, most certainly must be considered as the grand governing cause under Providence, of that animated spirit for trade and commerce, so prevalent in this kingdom; but the neglect or inattention to inferior causes, might eventually be highly prejudicial.

† The instances are so numerous which might be produced, of the beneficial effects of machinery, in improving our manufactures, that it is difficult to make a selection. A just and appropriate tribute of approbation might be given to *Sir M. Bolton*, and his *distinguished scientific partner, Mr. Watt*, as *well as to others;* but the following is selected as what perhaps may be deemed particularly striking.

the period when the commercial treaty with France was under consideration, some of the principal Woollen Manufacturers were alarmed by the danger our trade would be exposed to, from the admission of French cloths into this kingdom, unless it was guarded by high protecting duties.

The event fully proved the alarm to be groundless, though the Legislature complied, in some respects, with what those Manufacturers deemed to be necessary, otherwise the advantage to this kingdom, by our exports so much exceeding our imports, par-

Not more than between two or three years past, *chip hats* were made *only* from the upper part of the straw, about six or eight inches from the point. Whether the idea originated in seeing some Leghorn chip very thin, or whether it was really an original thought, the writer does not know, but an inferior mechanic, whose name he wishes to obtain, prepared a very simple apparatus for *splitting* the straw into six or more parts. The consequence was, that the *whole of the stalk* became capable of being used, and the demand increased in *an astonishing degree*, (by the improvement *in lightness*) not only for home consumption, but for exportation.

ticularly in inferior cloths, would have been more evident than *it actually was*.

In confequence of the particular fituation in which the writer of thefe Obfervations was at that time placed, he was not only enabled to procure fpecimens of all the Wollen Manufactures of France, but alfo to afcertain the prices given in the various proceffes of manufacturing goods.

From the information obtained, it was evident, though the nominal price of labour was cheaper in France, yet goods of equal comparative qualities were manufactured much cheaper in this kingdom.

How far the relative price of labour in the two nations may differ at prefent, from what it was when the commercial treaty was figned, is unknown to the writer.

That other nations, and more particularly France, will be difpofed to avail themfelves of the improvements in machinery, which for fome years paft have been effected in the Woollen Manufacture of this kingdom, cannot poffibly be doubted.

To place our fecurity on prohibitory laws, however rigidly enforced for preventing machinery from being carried out of the kingdom, would be fallacious; fufficient temptations will be held out to thofe perfons difpofed to incur the *risk of discovery*, for the *hope of gain*.

There is, however, no neceffity to rest on hypothetical reafoning on this fubject, as there can be fcarcely any doubt that nearly all the different kinds of machines, *at present* in *general use*, in this kingdom, either in one place or other in Europe or America, though poffibly imperfect in their conſtruction, are, at preſent, actually employed*.

* When the union with Ireland was under the serious consideration of the Legislature, the writer of these Observations was led to investigate a subject intimately connected with one which formerly he had attended to in a particular manner, but which never underwent that *full* and *impartial investigation* it was his wish to excite.

A favourable opportunity appeared to publish his renewed sentiments on the subject; but such reasons were assigned (though consistent with the enjoyment of the most perfect freedom of choice) by a highly respectable person, now filling one of the first offices in Government, that the making known the writer's sentiments, at *that time*, might be inexpedient; as he trusts will ever operate in inducing him to sacrifice his own private opinion, though not his principles, to what may, by respectable and competent judges, be likely, at the moment, to impede beneficial designs for *public good*.

One part of his subject led him particularly to wish for information respecting the knowledge possessed in France, respecting our improved machines, in the woollen and cotton trades.

Accidentally a friend of his happened to be in possession of a work published in that kingdom, in which were delineated on plates (with particular descriptions) all the machines used by the French, in the *woollen*, *cotton*, and *silk manufactories, in the year* 1758.

The inspection of this work fully convinced him it was not *merely through ignorance* of improved machinery that the

That the Legislature of this kingdom could be prevailed on to pass an act for preventing generally the use of particular kinds of machinery *already used* in the Woollen Manufacture, cannot be admitted as *possible to happen*, even though some partial evils may be proved to exist, in consequence of their introduction.

Still it is possible that reasons though in *themselves fallacious*, may, by men of real understanding, be assigned for not permitting a *further extension*, and it is within the bounds of probability, that if these reasons are not demonstrated to be erroneous, they may have considerable influence on the Legislature, from the hope of calming the minds of the work people, at the present moment.

French Manufacturers had not applied the carding engine, &c. to the woollen manufacture, but that it arose from other causes, which cannot, with propriety, be now noticed.

The writer, my Lord, is fully ſenſible that the application of the Clothiers, in the firſt inſtance, to the Houſe of Commons, will be ſupported by Members amply converſant in trade and commerce; but he preſumes to hope, he may, without incurring the imputation of arrogance, point out what in his opinion will be the conſequence of checking, *in any degree,* the ſpirit of improvement, not only to *the Manufacturers,* but alſo to the *work people themselves,* and to *the nation at large.*

The preponderating advantages poſſeſſed by this kingdom, in point of manufactures, may poſſibly prevent thoſe latent evils from being generally apparent, even for ſome conſiderable time, which the adoption of reſtrictive meaſures muſt *eventually* produce.

They will, however, *appear* in all their magnitude, the inſtant that rival nations ſhall attain to *an equal degree* of perfection with us in the *use* of machinery.

The lower price of labour, *though of itself* not ſufficient to excite uneaſineſs, when combined with other obvious cauſes, would then be *powerfully influential.*

This point once gained, the ſpirit of improvement would animate our rivals to attain *higher degrees of perfection,* while our manufacturers, checked by injudicious reſtrictions, would remain in *a torpid state.*

The neceſſary conſequence of ſuch meaſures muſt, to the *Clothiers,* be the total loſs of that part of their buſineſs which depends on the foreign trade.

When probably it might be too late to hope for redreſs, the *work people* would alſo find, that in conſequence of ſucceeding in their application for preventing the *con-*

tinual improvement of machinery, they had, though undeſignedly, been the cauſe of the *total loss* of our woollen *export trade*, and they would then actually experience that loſs of employment which they had *fondly hoped* the *prohibitory laws* would have prevented.

The *nation in general*, in ſuch a caſe, muſt not only ſuffer the loſs of a valuable part of its foreign trade, connected with all its relative conſequences, but muſt be burdened, at leaſt, for ſome time, with providing for the ſupport of thoſe perſons thrown out of employment.

Let this mode of reaſoning, my Lord, be applied to *all* other branches of manufacture, if a *general prohibition* was to take place, in reſtraining *progressive improvements* in machinery, what *incalculable evils* muſt *necessarily follow*?

Diſcuſſions reſpecting other manufactures, except ſo far as are intimately connected with the preſent ſubject, have been, and will be, particularly avoided, as the mind, fixed to one object, is more capable of forming correct opinions, than it would be by being attracted to ſeveral; the impolicy of impoſing reſtrictions reſpecting machinery will, therefore, except ſo far as *abſolutely connected* with any other, be confined entirely to the *woollen trade.*

In the further proſecution of the ſubject, the arguments uſed for proving the pernicious tendency of introducing machinery into the *clothing buſineſs,* may be comprised under the three following heads:

1ſt, It has been frequently objected by intelligent perſons, and even by ſome Manufacturers themſelves. that the introduction of machinery into the clothing buſineſs, muſt be pernicious, as only a *limited quantity* of the ſtaple article, wool, of our native growth, could be produced for our middling and coarſer manufactures, and that the poſſibility of obtaining an increaſed quantity of fine wool of the growth of Spain, for our ſuperior cloths, muſt depend on the demand for that wool from France, Holland, &c.

The concluſion drawn from theſe *supposed facts* is, that machinery muſt be prejudicial, by diminiſhing the labour neceſſary for the employment of a great number of perſons in making goods, without the poſſibility, as we do not, in the opinion of the objectors to the uſe of machinery, poſſeſs the means of *increasing the manufacture,* but in a very limited degree, even ſhould the *demand require it,* conſequently thoſe perſons could be no longer employed in the woollen buſineſs.

2dly, The corruption of the morals of the people, eſpecially of the children employed in the factories, in conſequence of collecting them together, has been confidered as a formidable objection to the ufe of machinery.

3dly, From the general introduction of machinery, it has been confidently faid, the ftaple manufacture of the kingdom is endangered, by its being fo much more eafily *transferred to other nations.*

The advocates for reftrictions affert that the ufe of machinery, by rendering the fuperior fkill of thofe perfons, long accuftomed to different branches of the manufacture, comparatively ufelefs, it will become perfectly eafy to form eftablifhments for the trade in other nations, as only a few fuperintending perfons will be wanted, who may be fully competent to the *management of machinery.*

In reply to the firft objection,

It may with confidence be affirmed the

aſſertion, that only *a certain quantity* either of our own *native wool,* or of *foreign growth,* can be obtained, is a mere vulgar error, originating at firſt either in ignorance or from deſign, and, through inattention, repeated by ſucceſſive writers in the various diſputes that have occaſionally occured on the ſubject of the *exportation of wool.*

It may, perhaps, be abſolutely impracticable to determine whether the growth of wool in this kingdom has *increased* or *diminished,* within the laſt thirty years.

When the Manufacturers applied for further reſtrictions to guard againſt the export of our wool, it was neceſſary to ſhew the value of the woollen trade to the nation. To effect this, the quantity of our native wool annually grown, was deſirable to be known; no other mode, however, occured than placing reliance on the opinion of *persons of experience,* who had, on ſome former occaſions, made calculations on the ſubject.

One of the reſpectable Yorkſhire Delegates, a conſiderable Woolſtapler, at the time when the union with Ireland was under conſideration, in his evidence to the Houſe of Commons, ſtated the number of packs to be 600,000.

This is the ſame number which was given to the Wool Meeting by a perſon of the ſame name, moſt probably the father of the perſon alluded to.

On what ground this eſtimate was *originally* formed, cannot be known; but though the writer always doubted whether ſuch a *very large quantity* was produced in England and Scotland, (for moſt certainly, at that time, Ireland could not be included) he had no means to aſcertain the contrary.

The refult of enquiries, made fome time paft in the county of Wilts, led to conclufions adverfe to a *general increase* of the growth of wool, but from information fince obtained from perfons refident in different parts of the kingdom, it feems probable that the number of fheep has actually increafed in *some counties**.

* The writer having had occasion to call on the respectable Member for Norfolk, who has so much distinguished himself, by his attention to agricultural pursuits, could not possibly avoid making some enquiries on the subject, that engaged his attention.

The particulars Mr. Coke was so obliging to communicate would, if mentioned, swell this note too much. It may, therefore, suffice to say, that by the introduction of the South Down breed of sheep, Mr. Coke has increased his number *very considerably indeed* on the *same tract of ground*, and that the fleeces are full as heavy as his former Norfolks.

This Mr. Coke stated to be pretty generally the case also with other gentlemen and farmers in the county of Norfolk.

The writer, from his partiality to Norfolk wool, had recommended to Mr. Coke, more than two years past, attention to the *county growth*, but it seems the advantages from the South Down sheep preponderate, in opposition to improving the Norfolk breed.

In the debate in the Houſe of Commons, at the time the union with Ireland was in agitation, when the Manufacturers petitioned againſt the exportation of wool to that kingdom, Mr. Pitt confidently aſſerted, that incloſures had been the cauſe of the *actual increase* of the breed of ſheep.

Though other cauſes may be aſſigned in oppoſition to this, of a contrary nature, or producing different effects, yet, whoever has peruſed the then Miniſter's ſpeech on that occaſion, with a mind diveſted of *all partiality to system*, will be diſpoſed to give him credit for *accuracy of information*, and a *thorough knowledge of the subject*, as far as poſſibly, from *prudential reasons*, he might at *that time* think proper to diſcuſs it*.

* Whatever *political*, or *prudential* considerations might influence Government, at the time the projected union with Ireland was under consideration, to allow of the *exportation* of British wool that kingdom, *free from any duty*, though *the duties on our woollen goods imported into Ireland*, were still to be continued, yet certainly, if the Act of Union *does not abso-*

The increaſed amount of our exports of woollen goods, as ſtated on the authority of Government, when the union with Ireland was under conſideration, will certainly afford reaſons to ſuppoſe the annual growth of our native wool has increaſed.

It muſt, however, be granted, that the amount of the value of the exports taken from the Cuſtom Houſe books; is by no means a proof of itſelf, that the *quantity* of goods exported, has actually increaſed.

The price of different kinds of woollen goods has really advanced of late years, and

lutely prevent it, time must arrive, when British woollen manufactures shall be *imported* into Ireland, as freely as the *raw material* is permitted to be exported to that kingdom from hence.

That the writer of these Observations ever rejoiced in promoting, as far as his limited situation would admit of, the real prosperity of Ireland, is evident, by the part he took in explaining objections against those parts of the *Irish propositions*, which were relative to the *woollen manufacture*, and for which, by written communication from Government, it was acknowledged he had *rendered service* to his native country.

ſhould this, on the whole, appear to be *any thing conſiderable*, the *nominal value* in the Cuſtom Houſe entries might have been much larger than in any former given period, though the quantity *might be actually lessened.*

The writer, my Lord, is conſcious that arguments ſimilar to thoſe he is now uſing, requiring ſome degree of attention, are free-quently neglected by the reader; but if he is not greatly miſtaken, they are of conſequence, not merely to the ſubject he is conſidering, but to the *general trade of the nation.*

He feels himſelf no anxiety for the trade of the kingdom, (though conſcious he cannot be indifferent to its welfare) if the energy of the Manufacturers, &c. is not cramped; but ſtill persons may be greatly deceived, by fixing their attention merely to the *increased value*, from Cuſtom Houſe entries, of the export of any article. This ſubject, though important in itſelf, muſt not, however, be further purſued at preſent.

To return, therefore, to the direct argument, respecting the probability of an increase in the annual product of wool.

For forming an accurate opinion on the subject, it would be necessary to know the *actual excess of the value of the raw material imported from Spain and other parts,* within the period that was given in the accounts laid before the House of Commons.

To this must be added its *additional value,* by *the expence incurred of manufacturing it.*

If this total amount should not appear to be equal to the *increased value of the exports* (making proper allowance for the advance in price, as noticed before,) in a certain period, it would afford a pretty decided proof, that our *native wool* was actually increased in quantity, except it should be supposed our *own internal consumption* of woollen goods *had been diminished**.

Though ſome allowance muſt be made for the advance in the price of goods, as before mentioned, yet perhaps ſufficient reliance may be placed on the account delivered to the Houſe of Commons, from the Cuſtom Houſe, to afford *general proof*, that our native wool muſt have rather increaſed in quantity.

In the year 1799, the exports of woollen goods, amounted to	£6,876,939	8	3
In the year 1790, they were only	5,190,637	13	6
Difference	£1,686,301,	14	9

* Though the use of Norwich and other woollen stuffs has so much declined, yet there is not the least probable ground to suppose our internal consumption of woollen goods has been lessened. Most assuredly people in general consume more

The account of the imports of foreign wools does not go back further than 1791, but one year can make no material difference in this general view of the ſubject.

In the year 1799, the total amount of the imports of Spaniſh and other foreign wools, was	4,935,839 lb. wgt.
In the year 1791, it was	2,776,54
Difference	£2,159,765

As it is impoſſible to aſcertain the proportion between the finer wools and others of the coarſer quality imported, there are no means of making an accurate calculation of the actual value.

The writer therefore conſiders it to be fully ſufficient to ſtate his opinion that the

cloths than formerly, and the increased use of carpets, &c. must require an additional quantity of wool.

original value of the wool, and its *increase* in price by manufacturing it, may be taken at about *one million two hundred thousand pounds sterling.*

The exceſs of the exports of woollen goods, in 1799, compared with 1791, has been given as,	£1,686,301	14	9
Calculating the value of foreign wool, and the expence of manufacturing it, as ſtated above .... Should be:	1,200,000, 0, 0 ~~1,200 0 0~~		
The difference would be	£486,301	14	9

That no *absolute* dependance can be placed on the Cuſtom Houſe entries, will be granted; but as it may be preſumed the ſame irregularities reſpecting the entries of an article paying no duty, ſubſiſted *equally in* 1791 *as in* 1799, though there may poſſibly be errors in *the total amount*, in both years, yet there can be no great difference in *the comparative view*.

From the above ſtatement it appears, the ſurplus of exported woollen goods, in 1799, after deducting the value of foreign wool imported, and its additional amount by manufacturing, over and above what it was in 1791, amounted to not much ſhort of *five hundred thousand pounds*.

If the premises are well founded, the inference is incontrovertible, that though *the import of foreign wool* was *so much increased*, from 1791 to 1799, yet *the export of goods* from *our own native growth of wool, greatly increased also**

* It would be improper to enter on the subject of wool, at present, only considered as connected particularly with the general objects treated of otherwise, *the policy* of the measure of *growing fine wool generally* in this kingdom, might be controverted.

The observations of Lord Somerville, respecting the probability of increasing the breed of sheep, in this kingdom, are

That this kingdom can be ſupplied with a much larger proportion of the fine wool of Spain, is undeniable, from the progreſſive increaſe of our *annual import*, and in proportion to the demand for fine wools, will the attention of the people of Spain be directed to the improvement of *their inferior wools*, as well as *a general increase of the article*?

In conſequence of the ſuperior price that can be given by our Manufacturers for wool, will our Merchants be enabled to import ſtill larger quantities of different kinds of wool from other parts of Europe?

certainly worthy of attention; and though his Lordship in his Address to the Board of Agriculture, was mistaken in some particulars in the application of his reasoning, and though some partial evils did arise at that time, from sanguine expectations of substituting fine English for Spanish wool, yet the Manufacturers, and the nation in general, will be benefited by the patriotic exertions of his Lordship, and other Noblemen and Gentlemen, to ameliorate our wool, *if attention be directed to the production of such kinds of wool, as the relative state of this kingdom, compared with other nations, respecting the growth*, may require.

From Africa the import of wool may be conſiderably increaſed, and our extenſive dominions in the Eaſt Indies, from the information the writer has obtained, and from ſpecimens of wool he has ſeen from thoſe parts, might, under proper management by the Company's Agents, be rendered tributary to our woollen manufacture.

When the woollen trade of France was in its moſt flouriſhing ſtate, different kinds of foreign wools, unknown in general to our Manufacturers, were uſed in that kingdom, and other ſorts of raw materials capable of being manufactured on *the same principles as wool,* were alſo made uſe of by the French Manufacturers*.

* Were it necessary, it might be shewn that the introduction of a particular kind of raw material into use for the woollen manufactures of this kingdom, between twenty and thirty years past, has excited a spirit of improvement, by the substitution of other articles, which, by degrees, may improve our fabrics to the advantage of individuals, and to the benefit of the nation in general.

From all theſe conſiderations, there does not therefore appear to be the leaſt reaſon for apprehending *any want of supply, of the raw material,* neceſſary for a very *considerable increase in the consumption,* and conſequently the objection to the further introduction of machinery, on the account of the *alledged limited quantity* of wool, *must be groundless**.

The ſecond objection to the introduction of machinery, is founded on the conſequent corruption of the morals of the work people, employed in the manufactory, eſpecially of the children.

This objection, in a limited degree, would have applied in all the different ſtages of improvement, in all arts and manufactures. It may be neceſſary previouſly to obſerve, that

* Some persons may possibly think a general answer to this objection, which might be given, amply sufficient, *That in proportion to the demand for any article, will be the supply,* but it was deemed expedient to be *more particular.*

the terms *manual labour*, and the *use of machinery*, are not ſo particularly conſidered as to be in general juſtly diſtinguiſhed, either in an abſolute or relative degree.

Strictly ſpeaking, that only is manual labour which a perſon can perform, merely by the operation of his own hands and feet, aided by other parts of his body, or by the perſonal aſſiſtance of one or more perſons, without the introduction of any machine whatever. For inſtance, one perſon might tear up roots with his hands, and ſcratch the ground with his nails; another might follow, and make furrows; a third might drop the ſeed; and a fourth might cover the ſeed with the earth ſcraped from the furrows; and this would be *actual manual labour*, employed in the cultivation of the ground.

Let us proceed one or two ſteps further, in the proceſs: one perſon digs the ground, by the help of a *spade*; another, by diſcovering the benefit of breaking the clots of earth, has contrived ſomething like *a harrow*, to draw over the ſurface of the ground; a third follows and ſcatters the ſeed; and a fourth finds the means, by *some instrument*, of covering it in a partial manner.

Viewing theſe operations, if the queſtion was aſked, by what means the ground was thus cultivated? Moſt perſons would reply, by *manual labour*; and yet it is certain, though we call a *spade* a tool, and a *harrow* a utenſil, yet in theſe operations there is an intermixture of *manual labour* and *machinery**

Confining attention principally to the woollen buſineſs, to demonſtrate what has been

* This may, among others, be adduced as a proof that habituated to suppose positive ideas, annexed to merely comparative language, few persons are led to investigate just principles, and therefore, on this subject, as well as in others, we are liable to deceive ourselves in our modes of reasoning, by *analysing terms*.

juſt aſſerted, *that the objection, in a limited degree, would apply in all the different stages of improvement,* let it be conſidered there muſt have been *a time,* when ſcribbling of wool, in what may now be called the old method, by means of the *scribbling horse,* muſt have been unknown, as well as the carding the wool afterwards by women.

At ſuch a period let us ſuppoſe ſome ingenious Manufacturer had diſcovered the method of making cards, and applying them to the ſcribbling of wool, he would, moſt probably, be deſirous of making the moſt advantage poſſible of this diſcovery, and would therefore employ perſons in his own houſe, to perform the operation; by degrees it would become generally known, and the utility of having ſeveral perſons together in one work ſhop, for facilitating the operation, would be evident.

In conſequence of this diſcovery, women and children, who before this time were employed in ſeparating the entangled filaments of the wool, perhaps by different *manual processes*, muſt have been deprived of ſuch employment. The facility, however, with which they could now form threads, in compariſon of what they could do before this *invention of scribbling*, would immediately reconcile them to it. The neceſſary conſequence of this firſt diſcovery, would tend to *carding the wool*, for preparing it in a ſtill better manner for ſpinning.

At ſuch a period as this, might not the objections ſtated againſt the *further introduction* of machinery have been urged, to ſuppreſs the *new-invented mode of scribbling*?

The men, it might have been ſaid, were to be taken from their families, and *being congregated in one place, would materi-*

ally corrupt each other's morals. The women and children would be thrown out of their immediate employment; and even ſuppoſing the whole number could now be employed in ſpinning, ſtill wool being of *limited growth,* a larger quantity than was uſed before *could not be procured.* The objection under conſideration is, however, confined to *the corruption of the morals of the work people,* and therefore the obſervations in reply muſt be confined more immediately to that object.

That it is peculiarly deſirable to keep both women and children, as much as poſſible employed at home, cannot be controverted.

At the time ſcribbling and carding the wool, by the uſe of more complicated and difficult machines began to be generally uſed, peculiarly happy would that perſon have been, who, perceiving the conſequences, could have influenced the Clothiers, *as a body*, to have at first confined the ſpinning of wool by jennies, to the exceſs of quantity, which the improved ſtate of the manufacture might have required.

By ſuch a reduction in the price of hand-ſpinning, abſolutely fixed between the Clothiers and the farmers, in the ſpinning diſtrict, as could with propriety be effected, by ſending the wool in an *improved carded* ſtate, the evil would have been greatly abated.

This expedient *was adopted* by many Clothiers, but, like all other *partial efforts*, by degrees failed, from its interfering too much with *private interest*.

Human eſtabliſhments, in the very nature of things, muſt conform to fluctuating circumſtances, attendant on human affairs, and thoſe meaſures are only truly wiſe, which have for their objeɑ *the mitigating* the evils which, from the imperfeɑion of all ſublunary things, are attendant on every kind of improvement. By the adoption of judicious meaſures, ſuch is the beneficent deſign of Providence, even thoſe very evils, in almoſt all caſes, by wiſdom and prudence, may be converted into *positive good.*

By improper management, and a total omiſſion of attention to the morals of the perſons employed in faɑories, no doubt, in ſome inſtances, they may preſent a ſcene in ſome conſiderable degree, to warrant the ſtrong language of a *Tourist,* some time paſt, into the northern parts: "That in ſome of the faɑories, the immorality and corruption of the work people exceeded any thing to be conceived of this ſide the infernal regions." This perſon, however, might, at preſent, if he choſe it, viſit ſome extenſive faɑories, where by ſtriɑ, though temperate diſcipline, attention to decency, and to the proper inſpection of the children's conduct, and alſo by judicious modes of managing the men and women, *the most perfect order is preserved.*

The Legiſlature has already interfered in the regulation of cotton factories, and, no doubt, equally beneficial, may be its interference for regulating thoſe in the woollen trade*

The objection, therefore, cannot juſtly be conſidered as *inevitably* militating againſt the

* Let any person view the order and regularity observed in the management of a great number of persons, in one of the many noble institutions in London, for the relief and the employment of persons in distress, and consider what might be the situation of those persons when they were separated, and he may feel his objections to factories, *supposing them to be properly regulated*, in some measure diminish. By this observation, the writer would not wish to be considered as an advocate for collecting, indiscriminately, a number of persons together, either for *employment* or for *parochial relief*. Some places, of each description, are absolutely necessary, but the strong links that attaches the poor to *their families*, *their friends*, and *their country*, should not be wantonly severed. The old English proverb is worthy of attention—Home *is home*, though ever *so homely!*

introduction of machinery, as it is evident the supposed ill consequences to the morals of the people may be obviated.

The third objection remains to be considered, That the adoption of machinery facilitates the transferring of the woollen manufacture to other nations.

This objection the writer has heard advanced, by sensible persons, who not only saw the necessity, but have actually adopted improved machinery, and therefore it deserves particular notice.

To consider the woollen manufacture exactly of the same consequence as it was, perhaps, justly esteemed by our ancestors, would lead to an erroneous conclusion, *as that on which almost the very existence of the nation depended.* It may, undoubtedly, be still denominated our grand staple manufacture, as the quantity of the raw material *imported,* though conſiderable in itſelf, bears ſo ſmall a proportion to the *native raw material,* uſed in manufacturing woollen goods.

The cotton manufacture, (to ſay nothing of others) is, however, ſo much increaſed of late years in this kingdom, that its intereſts demand particular attention, as well as the woollen manufacture.

The introduction of improved machinery, originated in the cotton buſineſs; and its very exiſtence, in this kingdom, depends perhaps not merely on *the continuation* of the preſent improvements, but alſo on *progressive improvement,* for enabling the Manufacturers to contend againſt the advantage from the lower price of labour, poſſeſſed by other nations. It is a fact, not to be diſputed, that improvements in mnchinery in the manufacturing of cotton goods, by judicious attention, may be applied to the woollen manufacture.

To guard, therefore, againſt the tranferring of our *woollen manufacture* to other nations, on *the principle advanced* by the advocates for reſtrictive meaſures, we muſt be reduced to the neceſſity of *limiting improvements* in the cotton buſineſs, which, in the courſe of time, would inevitably *prove its destruction.*

Though it has been admitted, that in ſome place or other on the continent of Europe, our improved machinery may, in *some degree,* be uſed, yet moſt certain it is, at preſent in its *infant state,* compared with our's, and the mode of managing it very imperfectly known.

While other nations may be combating with the difficulties attendant on the firſt introduction of improved machinery, and poſſibly with ſimilar prejudices from the work people, at firſt experienced in this kingdom, and requiring a ſeries of years to arrive at the point we have *at present at-*

tained, this kingdom may be progreſſively advancing in its career; and ſhould they ever attain to our *preſent ſtate of improvement,* we ſhould, at that period, poſſibly be paſſed beyond them in increaſing ſuperiority, *juſt as far,* if *not farther,* in comparative excellence, than we are at this preſent moment of time*.

The true ſecret, therefore, for retaining our manufactures muſt be ſought for, not in *reſtrictions on the uſe of new machines in manufactories,* by which the *efforts of inge-*

* Since writing the above, authentic information has convinced the writer that particular application is *now* making, by persons who have left this country and settled in another kingdom, for procuring men capable of managing machinery in the woollen business. He is however equally convinced, by the same information, that a *very small progress* is made there as yet, in machinery, and therefore, as it is absolutely impossible *wholly* to prevent the emigration of useful persons (though every means, consistent with the principles of natural equity should be used to prevent it), or the transportation of our machines to foreign nations, that mode of reasoning appears to be most conducive, which inforces the necessity of *progressive improvements,* instead of *severe restrictions.*

nious men may be paralized, neither in *contracted regulations,* for the *supposed benefit of trade,* or in *harrassing and injudicious regulations*—but in *unfettered improvements,* in *the enlightening of the minds* of the *work people,* to *discover their true interest,* in the *repeal of obsolete statutes,* wholly *inapplicable to the present state of the business,* ſerving only for a pretext to *interested men to deceive the people*; and, laſtly, in freedom from *oppression* and *injudicious taxes.*

Having thus endeavoured, with impartiality, to obviate the moſt material objections to the introduction of improved machinery into the woollen trade, it may not be improper to add a few miſcellaneous obſervations, applicable to the general ſubject.

The Gentlemen deputed from the Clothiers, to attend the progreſs of the Appeal to the Legiſlature, for repealing the obſolete ſtatutes, will, no doubt, make ſuch particular obſervations on them, as may enable the Houſe of Lords and the Houſe of Commons to diſcover the true reaſons on which their repeal is repreſented *as necessary.*

It is, therefore, not the intention of this Addreſs to enter upon a *formal examination* of all acts relative to the woollen trade; but by curſory remarks, to excite attention in their diſcuſſion by the Legiſlature.

Poſſibly the exiſting ſtatutes relative to the woollen trade, may be conſidered under the following diviſions:

1ſt, Such as are abſolutely impoſſible in the preſent ſtate of the trade to be complied with.

2dly, Such as not only militate againſt the neceſſary alterations that have taken place, in compliance with the prevalent ſtate of the trade, but may alſo be conſidered as hoſtile to improvements in machinery.

3dly, Such as may be beneficial in the general principle, but which require attention to adapt them to the *improved state* of the woollen manufactory.

Under the 1ſt head may be compriſed ſeveral ſtatutes reſpecting the mode of dying wool and cloth, and alſo the preſſing of cloth.

Under the 2d, Such as relate to regulations in the breadth and length, as alſo the weight of different ſorts of goods*.

* The regulations for the breadth as also for the length of those cloths, sold by the piece, were certainly proper at the time they were made, and no doubt the foreign trade has been injured by the debasement in quality, and by contracting breadths, &c.; and the writer could make this apparent, if necessary, from information he received many years past, from a respectable Turkey merchant.

Such as are relative to tenters and the ſtretching of cloth; and ſuch, particularly, as either directly or indirectly apply to the dreſſing of cloth.

Under the 3d head may be included,

The ſtatutes of Ed. III. and IV. prohibiting the importation of foreign woollen goods.

The ſtatute of Ed. IV. prohibiting the importation of *undressed* cloth.

The ſtatute of Anne, for impoſing a duty on *undressed* cloths; and all thoſe ſtatutes enacting particular regulations for the trade in Yorkſhire, which include the appointment of inſpectors.

The French, at one time, were particularly careful in the regulations for their export woollen trade; and some of their most judicious writers consider *its decline* as consequent on *inattention to these regulations.* Though it may therefore be necessary to make *some alterations,* yet it is equally necessary to guard against an *injudicious application* of a *general rule.*

And laſtly, Such ſtatutes reſpecting apprentices, as are applicable to the woollen manufactory.

The grand and leading point, exciting the oppoſition of the workmen, in the county of Wilts, has been the application of the machine, known by the name of Gig, to the dreſſing of fine white as well as medley cloth; and this oppoſition they deem juſtifiable, by the ſtatute of the 5th and 6th Ed. VI.; and as the writer has, perhaps, conſidered the legality of its uſe, in rather a different point of view from the Manufacturers in general, he is diſpoſed to conſider its hiſtory particularly.

The machine called a Gig Mill, has undoubtedly been uſed in Glouceſterſhire as well as in Wiltſhire, for dreſſing *coarse white* cloth, it is preſumed longer than any perſon can remember. How *long* it has been applied in Glouceſterſhire to the dreſing of *fine white* goods, is unknown to the writer of theſe Obſervations.

Whether the Gig Mills, *now used,* can be conſidered as exactly the ſame as thoſe deſcribed in the ſtatute of Ed. VI. may admit of *reasonable doubts.* The preamble to the ſtatute is as follows:

"For as much as true drapery of woollen cloth is to be commended, as well in foreign parts as in the realm of England, and all ways and means uſed to the contrary, are to be eſchewed and taken away; and for as much as, in many parts of this kingdom, is newly and lately deviſed, erected and builded and uſed, certain mills, called Gig Mills, for *perching* and *burling* of cloth, by reaſon whereof the true dreſſing of the cloth of this realm is *wonderfully impaired,* and the cloth thereof *deceitfully made,* by reaſon of the uſe of the ſaid Gig Mills, for the remedy thereof be it enacted, &c. &c."

Burn takes no notice of this ſtatute, and poſſibly a minute examination may diſcover its repeal. If the preſent *Gig Mills* are ſimilar to thoſe deſcribed in the ſtatute, either they muſt have undergone ſome alteration, or elſe their uſe muſt, from *their utility*, have been so apparent as to be connived at; this may reaſonably be concluded from their having been continued to be made uſe of in the dreſſing of *coarse* white cloth: this has been the caſe, even in the *county of Wilts*, longer than any one can remember.

However this point may be determined, most certain it is, that the preſent mode of dreſſing fine, as well as coarſe cloth, by *Gig Mills*, ſo far from *wonderfully impairing the true drapery of this realm*, and *the cloth thereof deceitfully made by reason of the use of the said Gig Mills*, that no maſter ſhearman of character, examined in a Court of Juſtice, would hazard *such an assertion*; and poſitive proof, ſufficiently ſatisfactory to convince the minds of an impartial jury, might be adduced by the Clothiers, to ſhew its *superiority* to the *common mode of dressing cloth.*

An inſpection of the ſtatute of Ed. IV. properly explained in a court of juſtice, might probably have a ſtrong tendency to convince the Court and a Jury, that what are *now called Gig Mills*, at leaſt as to the *principle of working*, were actually in uſe ſo long ago as the reign of Ed. IV.

By the ſtatute of the IVth of this King's reign, it is enacted as follows: "Every *fuller*, in his craft and occupation of *fulling, rowing*, or *tayselling of cloth*, ſhall uſe *taysels*, and *no cards*, deceitfully impairing the cloth, on pain to yield to the party grieved his double damage."

By this ſtatute it is evident, the fuller, or *millman*, at that period, actually *dressed cloth*; and the terms, rowing or tayſelling of cloth, in the ſtatute, are more accurately deſcriptive of the uſe to which the preſent *Gig Mills* are appropriated, than the terms *perching* and *burling*, applied by the ſtatute of Ed. VI.*

All reſtrictive ſtatutes are, by the liberal ſpirit of the common law of this realm, required to be conſtrued *strictly*, as they

* The information the writer has received in conversation with a manufacturer from Gloucestershire, whose father was a *fuller of cloth*, since writing the above, seems to confirm this mode of reasoning. This person informed him, that, to the best of his recollection, there is scarcely a fulling mill in their neighbourhood, but in which there is at present a Gig Mill, most of them apparently of old construction; and even some time past, in some of the old mills, where no *Gigs* were *then* in use, it was evident they had been used by the construction of the mills for that particular purpose. That the Gig Mills have continued in use for a very long period of time, is evident, from the terms used in Gloucestershire, by those who work them—*Varming the Mill*, equivalent to *setting it*—*Runge the Mill*, for *turning the Mill*, are still used. The person who

deprive the ſubject of ſome of his original rights. The only juſtifiable reaſon the common law admits of, for the reſtriction of the right to do any thing, *not contrary to what it admits* is *public benefit.*

Though it is, by no means, deſigned to reſt the validity of theſe Obſervations on the *construction* of any obſolete ſtatute, yet it is of material conſequence for the work people to be warned againſt the ſophiſtical reaſons aſſigned by their intereſted adviſers.

To ſuppoſe that any manufacturer could be ſubjected to the penalties inflicted by the ſtatute of Ed. VI. for uſing the preſent *Gig Mills* for dreſſing of cloth, would be paying a very bad compliment to the enlightened judgment of the courts of juſtice, *if due care was taken in explaining the subject.*

gave the writer this information, says, the terms *Varming* and *Runge*, are confined in their use, *entirely* to *managing the Gigs.*

The preamble to that ſtatute ſtates, *explicitly*, the grounds on which it was enacted by the Legiſlature.

Whether the arguments at *that time* uſed by the oppoſers of the Gig Mills, were juſt or not, cannot be material to the preſent purpoſe, though probably they poſſeſſed no more *real force* than thoſe now uſed.

It is moſt undeniable that no *strict proof can be adduced,* that what was called a Gig Mill in the reign of Edward VI. was ſimilar to thoſe *now* in uſe, (though the name is retained) as one of the purpoſes thoſe Mills were ſaid to be applied to, viz. the burling of the cloth, is not even performed by thoſe now uſed.

Can it therefore poſſibly be ſuppoſed, for uſing a machine not coming within the deſcription of thoſe proſcribed in the reign of

Edward VI. and which, ſo far from *wonderfully impairing* the cloth of this realm, *actually improves it* to *a very high degree,* and *thus commends it to foreign parts,* a perſon ſhall be liable to the penalties inflicted by the ſtatute of Edward VI.?

If this cannot be admitted, ought not the ſhearmen and others concerned in oppoſing the uſe of the Gig Mills, duly to conſider, that even were no application to be made for the repeal of this ſtatute, it would be conſidered as *virtually repealed?* and conſequently the *very ground* on which they reſt *the justice of their cause, is not tenable.*

With the *moderate* and *well-disposed* part of the work people, it may be hoped this argument will have ſome weight, as at preſent they may conclude they are only contending for the ſuppreſſion of what the *law itself* conſiders to be a great *national evil.*

Though the work people, in the county of Gloucefter, have, as the writer fuppofes, joined thofe of Wiltfhire and Somerfetfhire, in the general principle of oppofition, yet it is fcarcely credible they fhould be advocates for the *destruction of Gig Mills*, as moft certainly the prefent fuperiority of the county generally confidered, as to finifhing goods, (and even as to dying in fome refpects) muft be attributed to the ufe of the Gig Mills.

If the work people in the counties of Wilts and Somerfet in particular, can be prevailed on to confider the fubject as it affects their *real permanent* interefts, they muft be convinced that, for the prefervation of the trade in the two counties, they really fhould be defirous the Clothiers may legally (fuppofing however what is not granted, that they can not do it at prefent,) avail themfelves of the ufe of the Gig Mills, as freely in the fine trade as they have in fact in many inftances in the coarſe trade, for a ſeries of years paſt.

The dictates of real wiſdom would therefore lead them to ſee their future welfare, conſiſts, not in oppoſing the intended meaſures of the Clothiers by renewed acts of aggreſſion, but in calmly ſtating to Parliament thoſe reaſons which lead them to ſuppoſe the alterations intended, will be prejudicial to their intereſts, and truſting, if alterations ſhould be found abſolutely neceſſary for the Clothiers, that the Legiſlature will adopt ſuch regulations in favour of the work people, as ſhall directly tend to *blend* and *unite* their *private interest* with the *public good.*

Such a line of conduct would not only procure for them reſpectable ſupport, at preſent, for guarding what, on free examination, may be found to be their eſſential rights, but would alſo inſure to them, in future, the means of redreſs, ſhould any *real* and *per-*

manent evils arife from fuch alterations in the laws, as the Clothiers may ultimately obtain.

It was ftated that it was not the defign of the writer of thefe Obfervations to confider all the points minutely, as thofe Gentlemen examined by the Committee of the Houfe of Commons, will, no doubt, direct their attention peculiarly to it; other remarks refpecting the utility of the Gig Mills, will therefore not be brought forward.

The fhearing frames, moft certainly, are not of that confequence to the *improvement of the trade,* as the ufe of the Gig Mills, and were the work people to difcover a difpofition to be governed by the dictates of reafon, it is to be hoped the Clothiers would not deem the introduction of the frames, as *indispensably necessary,* in any greater degree than at prefent, till it might be evident the ſhearmen could, in ſome way or other, be fully employed.

No agreement certainly can be entered into for the ſhearing frames *not being used,* as this would militate againſt the principle which has been ſhewn to be beneficial in its operation, not only to the general trade of the kingdom, but to *the work people themselves;* it is not, however, to be ſuppoſed, but that a return to order and decorum, on the part of the ſhearmen, would be productive of thoſe liberal ſentiments, on the part of the Clothiers, which they appear to have manifeſted towards the ſhearmen, ſome time *previous to the riots.*

For the reaſon before aſſigned, thoſe ſtatutes reſpecting apprentices, which apply to the woollen manufactory, will not be minutely examined.

That ſome of them, if rigidly enforced, would be injurious to *the work people*, muſt be evident *to themselves*, if they would allow their own reaſoning powers to operate, uninfluenced by the ſuggeſtions of intereſted perſons*.

* The conduct of the weavers in Spitalfields, at the present moment, strongly illustrates the assertion, that a strict adherence to the statutes would be *injurious to the weavers*.

The silk trade, for some time past, having been much brisker than usual, a sufficient number of hands could not be found to *wind* the silk, owing to so many of the young women having taken to weaving.

It was suggested to the weavers, by some of their employers, that by preventing any person not having served a regular apprenticeship, (or not having been a weaver full seven years,) from being employed in weaving, would remedy this evil, by obliging those who now were weavers, to take to their former employment of *winding* the silk.

This project, *at first*, seemed to please the weavers, but, on mature consideration, they found, were it to be adopted, *their own children*, in comparison with others, would be the greatest sufferers; the workmen, therefore, *wisely* chose to bear *a less evil*, rather than incur *the risk of a greater*.

This fact, which has come to the writer's knowledge, since he began these Observations, would afford him an opportunity

The following Obſervations, taken from a caſe cited in *Burn's Justice,* are, however, ſo applicable to the preſent diſpute, that a recital of them may not be uſeleſs.

" By the common law, no man may be " prohibited to work in *any lawful trade,* " or in more trades than one, at his pleaſure. " So that without an act of Parliament, no " man may be reſtrained either to work in any " lawful trade, or to uſe diverſe myſteries " or trades; therefore an act of Parliament, " made to reſtrain any perſon herein, muſt " be taken ſtrictly, and not favourably, as " acts made in *affirmation* of the common " law.

There was an ancient ſtatute, 37th Ed. III. " That artificers or handicraftſmen, ſhould

to point out more particularly the beneficial consequences to be expected from *improved machinery* in general; this, however, would not be compatible with his *present design* of confining attention to the *woollen trade* principally.

" uſe but one *mystery*, and that none ſhould " uſe any myſtery, but that which he had " before that time choſen and followed. " But this reſtraint of trade and traffic, was " immediately found prejudicial to the pub- " lic, and therefore, at the next Parliament, " it was enacted, that all people ſhould be " as free as they were at any time before the " ſaid ordinance."

And Lord Coke obſerves, that " Acts of " Parliament, made againſt *the freedom of* " *trade*, never live long."

Two remarks may be made on the above extract:

1ſt, That the Legiſlature, from *partial* and *interested* application, may impoſe ſuch reſtraints on the exerciſe of trade, as may be *prejudicial to its real interest.*

2dly, That as one of the brighteſt luminaries of the law, expreſſed his diſapprobation of reſtrictions on the freedom of the Trade, they ſhould never be ſanctioned by the Legiſlature, except *in very peculiar cases,* which may juſtly be conſidered as *exceptions to the general rule.*

Indiſcriminately to arraign the wiſdom of our anceſtors in requiring a long apprenticeſhip in all trades, might juſtly be deemed raſh and preſumptuous. It does not, however, follow, that regulations adopted in the *infancy* of *trade and commerce,* or even in their *progress to a comparative state of perfection,* however juſt and proper they might be when eſtabliſhed, are never to be altered in conformity to exiſting circumſtances.

Granting, in the fulleſt extent poſſible, that the ſtatutes reſpecting *Apprenticeships* were in every reſpect conſonant to the dictates of wiſdom, when originally enacted, yet, undoubtedly, an alteration in the circumſtances relative to trade and commerce, may require, at preſent, ſome material alterations to be made in them.

To conſider *all* regulations reſpecting *Apprenticeships*, or of *trade in general*, as emanating *solely* from the wiſdom of the Legiſlature, would manifeſt very imperfect and confined ideas on the ſubject.

When Sir Robert Walpole ſaid, " The Merchants and Manufacturers were a ſet of ſturdy beggars," he might uſe *an obnoxious* and *impolitic phrase*, but it would require no great penetration to point out many caſes, where *importunity* and *self-interest*, by exhibiting ſpecious arguments, have influenced the Legiſlature to adopt meaſures reſpecting trade and commerce, *obnoxious to the general interest of the nation.*

When particular occupations began to be ſeparated, and to be carried on by ſeparate and diſtinct claſſes of perſons, they became *objects of trade* to *those persons*, and it was natural enough for them to obtain certain privileges, and to conſider the occupation as *a certain craft* or *mystery*, which required long experience to be maſter of*.

How far theſe cauſes might contribute, in influencing the Legiſlature to eſtabliſh the general regulations reſpecting apprenticeſhips, it may be difficult to determine; but it cannot be deemed an unreaſonable ſuppoſition, to conſider them as having *some* influence.

To what cauſe can it be aſſigned, that no *Apprenticeships* have ever been deemed neceſſary for enabling a perſon to act as a *Master*

* The policy of the Monarchs, in granting peculiar privileges to persons disposed to congregate in towns, by which it was hoped to weaken the power of the *great Barons* of those times, no doubt contributed to sanction the generally received opinion respecting *crafts* and *mysteries* in trade.

or as a Workman, in any of the *employments in husbandry?*

Evidently for this ſimple reaſon, that agriculture has remained in its *original primitive state.*

Had agriculture ever been diſtinctly and abſolutely ſeparated, for inſtance—into three diviſions, the *Dairy—Grazing* or *rearing of Cattle—cultivating the Soil* for the *production of Corn, &c.* it might have been deemed as neceſſary for a perſon to ſerve an *Apprenticeship* to qualify himſelf to be a *good Dairyman*, a *prudent Grazier*, or a *skilful Cultivator of the Soil*, as it was to be *an expert Shoemaker*, or *an adroit Taylor*, &c.

Let it not be ſuppoſed, by this mode of reaſoning, it is deſigned to inſinuate, that in all caſes, *Apprenticeships* to particular trades, are to be conſidered, *in themselves*, as real evils—*but merely to trace some* of the caufes that might operate in their efta-blifhment.

Sanctioned by *the opinion of Lord Coke,* it ftill may be fafely affirmed, that if from a change of circumftances, any reftricting fta-tute, fuch as that for inftance, refpecting *Weavers* in the *Woollen Business* fhould be found, in the courfe of time, *to be injurious to the freedom of trade,* it ought to be abo-lifhed.

The fame facred principle however which prevents any perfon's private property from being facrificed to public good, without his receiving an adequate recompence, fhould operate alfo in all cafes of the kind now under confideration.

If the general welfare of the *Woollen Trade* requires an abfolute repeal of the ftatute— by which no perfon, not having ferved a *regular apprenticeship of seven years,* or who has *actually worked seven years as a weaver,* ſhall be permitted to weave broad cloths, &c. no doubt the future intereſt of of the parties affected by the repeal ſhould be duly regarded*.

* That the Weavers are misled by false reports is certainly evident, by a particular fact in the writer's knowledge. He had been informed, that a considerable Manufacturer had declared, as soon as the bill passed, he would introduce weaving into his factory, and oblige the persons to *weave on his own terms.*

Though *a person of credit* stated this to be what was declared in the county to be a fact, yet the writer intimated the improbability of the declaration, even on the ground *of policy merely.*

Happening to meet the Gentleman alluded to* a few days afterwards, who was just come from the country, the writer did not scruple to mention the report, though observing he gave no credit to it himself. In reply, the Gentleman assured the writer, in the strongest manner, that no such declaration *had ever been made by him,* nor had he any *such intention.*

The writer could produce *decisive proofs,* that he himself never could be disposed to approve of any mode of proceeding, likely *to injure the work people;* on the contrary, he may safely appeal to the testimony of those he has employed, that he ever wished to *see them comfortable;* but in publishing his sentiments on the subject under consideration, he considers himself

* J. Jones, Esq.

In like manner the future intereſt of the *Shearmen* ſhould be reſpected, if it be found neceſſary to make alterations in the regulations preſcribed in any ſtatute or ſtatutes reſpecting *Apprenticeships*, in that branch of manufacture*.

In the rapid improvements making in all our manufactories by the introduction of new machinery, it ſurely cannot be beneath the notice of the Legiſlature itſelf to interfere as far as it may be found *consistent with the freedom of trade*, to guard againſt the evils ariſing from *sudden alterations*. This may be more eſpecially expedient when the inter-

bound, on principles of *justice and equity*, as it is *in his power so to do*, not only to *contradict an assertion* so injurious to the person alluded to, but to express his firm belief, that nothing *really* injurious to the Weavers is intended.

* Though the cases are not analagous, yet the remark, the writer understands, has been made in the county of Wilts, that the arguments used in the Courts of Law, respecting copy-rights in an author, may be applied to the case of the Shearmen and Weavers —is not *wholly irrelevant*. Most assuredly they should not voluntarily be *made sufferers*, to promote *general good*, without some adequate recompense.

eſts *of large bodies of work people* are concerned.

It is not ſufficient to ſay, every thing will find *its proper level*; and if the people are *deprived of one particular mode of employment,* they will *find another!*

This argument, in the courſe of theſe observations, has been admitted, *in its fullest extent,* as what will *eventually* happen, ſuppoſing *nothing adverse occurs* to prevent the operation of general cauſes; but every perſon of conſideration will be diſpoſed to pay due regard to the *immediate* interests of thoſe perſons, who may be materially affected by the propoſed alterations in the preſent laws.

At a moment like the preſent, when the united force of the whole nation may be requiſite to oppoſe the hoſtile deſigns of thofe, who, from unfortunate caufes, may become our open and avowed enemies, moft certainly, in the adoption of neceffary meafures, though they may be ultimately for the benefit of trade *in general,* peculiar attention is due to the interefts of the work people, for preventing *even temporary evils,* as far as it can be effected, though they may manifeft fome oppofition to the intended alterations*.

Thus, my Lord, has the writer endeavoured to consider the fubject, under every point of view which appeared to him to be neceffary, as it may be fuppofed to affect

* Should the Legislature, at any time, have it in contemplation to enact a statute, favourable to the cotton trade, but which might, in its effects, be deemed particularly injurious to the woollen business, would not all the *Woollen Manufacturers* unite in a body to oppose the bill, on its first entry to the House of Commons.

It is, therefore, not the *act of opposition itself,* but the *manner in which it is conducted,* that must discriminate it, either as *criminal,* or merely *defensive of supposed interests.*

the *interest of the Manufacturers*, the *welfare of the work people*, the profperity of the *county of Wilts in particular*, and *the trade of the nation in general.*

He has declined availing himfelf of any information he might have obtained from procuring an infpection of the printed obfervations, defigned for the information of the Select Committee, or even of the Report itfelf, not only out of refpect to the Gentlemen delegated to manage the bufinefs, but that the fentiments conveyed through the medium of this Addrefs, fhould emanate entirely from his own ideas*.

* The destructive tendency of private regulations among the work people, in different branches of trade, not only in checking improvements by the masters, but also operating to the injury *eventually* of the work people themselves, would open a wide field for investigation; and the writer is in possession of several prominent facts to warrant the above observation; but though, no doubt, it may be deemed to be connected, in some measure, with the subject under consideration, yet it is not absolutely *comprised* in it; and therefore, though of considerable importance in itself, *it is at present* declined.

In treating the general ſubject, the writer has unavoidably been led, in moſt caſes, to adopt the *general terms* uſed in writing on *trade* and *commerce.* For inſtance, in drawing a comparative view of the ſtate of the counties of Wilts and York, at a particular period, he has ſpoken of the Manufacturers of the latter county as the *rivals* to thoſe of the former, but though he cannot avoid feeling a partiality for his *native county*, yet, moſt aſſuredly, he rejoices in the proſperity of the county of York, and ſhall conſider himſelf, at all times, happy to be in the leaſt degree inſtrumental in doing juſtice to the *spirited exertions of the Manufacturers of that county.*

In a more extended point of view alſo, *foreign nations* muſt be conſidered as *our rivals* in trade and commerce; but how happy would it be for the world at large, if, inſtead of the prevalent ſpirit of monopoly in trade, more generous and liberal views ſhould prevail.

Every nation, no doubt, is bound to improve all its natural and acquired advantages to the utmoſt; but it does not follow from this, that its true intereſt conſiſts in continually exerting *hostile acts* againſt the trade and commerce of others.

In moſt kingdoms there are ſome particular articles more congenial to climate, &c. than may be produced in other parts, and ſome particular ſpecies of manufactures are more adapted to one place than another.

If, inſtead of counteracting what ſeems to be the general beneficent deſign of the Author of Nature, theſe peculiarities were perfectly attended to, in the commercial intercourſe of nations with each other, they might, inſtead of exciting to acts of aggreſſion, be improved to promote *mutual prosperity**.

* Supposing all the differences existing at present between this kingdom and France happily terminated, and a new commercial treaty to be framed, would it not be consistent with

These caſual concluding remarks, it is to be hoped, will be pardoned, though not ſtrictly connected with the ſubject.

Whatever circumſtances the writer may be placed in, he truſts, in union with the moſt *genuine* and *sincere* deſire to promote the welfare of the nation to which he belongs, he ſhall ever experience that ſpirit which impels the mind to ſeek for gratification in *general happiness*.

He begs leave to ſubſcribe himſelf,

With ſincere reſpect,

My LORD,

Your Lordſhip's moſt obedient

Humble Servant,

JOHN ANSTIE.

May 8, 1803.

true policy to admit the importation of French *cambrics* freely, in order to procure the admission of any article or manufacture we might select as particularly desirable to be imported to France?

POSTSCRIPT.

SINCE the foregoing Obſervations were ſent to the Preſs, a reſpectable Clergyman, in Devonſhire, nearly connected with the the writer, who happened to be in town, on being informed of the propoſed publication, expreſſed his earneſt deſire that ſome alteration might be ſuggeſted, for removing difficulties in the Acts, relative to "Embezzlements, &c." in the Woollen Manufactory; as in acting as a Magiſtrate, he experienced conſiderable difficulty, in knowing how to proceed.

One inſtance he mentioned in particular, in which two Acts claſhed with each other, and, in the diſcharge of his duty, his humanity had to combat with his due regard to the Statute, under which, as the caſe ſtood, he *must have proceeded.*

The Writer's time will not admit of his devoting it, at preſent, to an attention to this ſubject; neither does it come immediately into his deſign; but he believes the perplexities felt by the Magiſtrates, in any of their proceedings in ſimilar caſes, originate in ſome of the Acts in Charles the Second's (and perhaps of other reigns), referring to Regulations for the Work People *not having been repealed,* when the laſt Act of his preſent Majeſty, relative to Frauds, &c. in the Woollen Trade, was obtained.

With due deference to the Legiſlature, it may be ſuggeſted, that the beſt mode of proceeding, reſpecting alterations in the old Statutes, referring to the Woollen Manufactory, may poſſibly be, by adopting and acting on the very judicious conſideration of Burn, in his excellent remarks on the *pos-*

sibility and *expediency* of reforming the Statute Law in general.

He ſays, ſixthly, " As to the reſt, to lay " all the Statutes and Clauſes of Statutes to- " gether, which relate to the ſame ſubject, " and out of the whole to compoſe *one, two,* " or *more uniform* and *consistent Statutes,* " and then to repeal all thoſe others, as work- " men *destroy the scaffolding,* when they " have *erected the building.*"

If thoſe Gentlemen who are deſirous of retaining all the old laws relative to the Woollen Buſineſs, as deeming them *absolutely connected* with *its preservation,* would give due attention to the concluſion of the paragraph juſt quoted from the *accurate* Burn, they would perceive, that even allowing them all poſſible merit, at the time they were enacted, moſt of them muſt be conſidered, in his language, *merely as scaffolding.*

It may ſurpriſe ſome perſons, but it is a fact, that *twenty-six* Acts of Parliament refer to *Apprentices*, and, poſſibly, ſome may have eſcaped attention.

The preſent enlightened age requires ſome material alterations, to be made in ſeveral of the laws relative to trade and commerce; and were the writer not restrained by the fear of its being ſuppoſed he was desirous the present addreſs ſhould be the vehicle of making known *personal injury*, he ſhould be tempted to ſtate a caſe reſpecting himſelf, as well as one of a ſimilar nature relative to a friend of his, demonstrative of the *necessity* of a reviſion of ſome particular laws.

In the courſe of his Obſervations, he might have been led from the very nature of the ſubject, and in ſome inſtances, where it would *even have elucidated it*, to refer more particularly to himſelf, than he has occaſionally ventured to do.

If any perſon should be diſpoſed to charge him with *Vanity*, his conſolation is, that however juſtly he appreciates, being noticed by perſons of diſtinguished rank and reſpectability, he is conſcious, the influence of *no passion whatever*, ever *has*, or *can in future*, tempt him to *sacrifice his principles*, either to *attain* or *to secure*, the favor of his ſuperiors.

With particular ſatisfaction he can now ſtate from the information he has obtained; that in proceſs of time, there is a poſſibility of obtaining any quantity of *fine* Wools that may be wanted for our increaſing Manufactures.

He will only add, that it not being possible for him to attend minutely to the correction of the preſs, some errors have been committed, which cannot now be rectified? and poſſibly ſome may eſcape his notice, in the annexed liſt.

FINIS.

ERRATA.

Page 3, line 14, *for* discussing, *read* examining.
—— 14, — 19, — aid, *read* injury.
—— 21, — 9, — Gloucester, *read* Gloucestershire.
—— 25, — 11, Note, *for* Sir M. Bolton, *read* M. Bolton, Esq.
—— 28, — 4, 5, — have been effected in the Woollen Manufacture, *read* have been introduced into the Woollen Manufactories.
—— 29, line 11, Notes, *for* that the making known, *read* that publishing.
—— 34, — 1, 2, *for* discussions respecting other Manufactures, except so far as are, *read* Observations respecting other Manufactures, except so far as they are.
—— 35, line 12, *after* without the possibility, *read* of their being employed in any other part of the trade.
—— 40, line 4, Notes, *after* British Wool, *read* to.
—— 41, — 1, —— —— prevent it, *read* the.
—— 45, — 11, *for* £. *read* lb. wgt.
—— — — 1, Notes, *for* cloths, *read* cloaths.
—— 47, — 3, —— a comma after treated of, should be inserted.
—— 52, last line Notes, the word not is omitted before *analysing terms.*
—— 62, line 11, —— *for* conducive, *read* conclusive.

Printed by C. Stower,
Charles Street, Hatton Garden.

CONSIDERATIONS

UPON A

BILL

NOW BEFORE PARLIAMENT, FOR REPEALING (IN SUBSTANCE) THE WHOLE

CODE OF LAWS

Respecting the

WOOLLEN MANUFACTURE

OF

GREAT BRITAIN:

And

FOR DISSOLVING THE ANCIENT SYSTEM OF

APPRENTICESHIP,

BY THE

Abrogation of the Laws relating thereto, as far as they respect the Clothing Trade, in certain Counties, mentioned in the said Bill.

LONDON:

PRINTED BY C. STOWER,

Charles Street, Hatton Garden.

1803.

Note for the reprint edition. The mistake in pagination—page 19 followed by page 28—occurs in the original printing.

CONSIDERATIONS

ON THE

BILL,

INTITULED,

" A Bill to repeal the Provisions, Regulations, and Restrictions, contained in certain Acts, passed in the 13th Year of the Reign of King Richard the Second, the 5th and 6th Years of the Reign of King Edward the Sixth, the 2d and 3d of the Reign of King Philip and Queen Mary, and the 4th and 5th Year of the same Reign, the 43d Year of the Reign of Queen Elizabeth, the 4th and 21st Years of the Reign of King James the First, the 10th Year of the Reign of Queen Anne, and the 1st and 13th Years of the Reign of King George the First, so far as the same relate to the Length, Breadth, and Weight of Woollen Cloths, the tentering, stretching and straining, and the viewing, searching and sealing of them by Officers appointed for that Purpose, the Boiling of Wool with certain Ingredients, the Prohibition of the Use of Lamb's Wool, the Sale of Woollen Cloths tacked and folded, the pressing of Woollen Cloth, and the Use of Gig Mills, and the Number of Looms; and also to Repeal an Act, passed in the 5th Year of the Reign of Queen Elizabeth, intituled, " An Act containing divers Orders for Artificers, " Labourers, Servants of Husbandry, and Apprentices," so far as the same relates to Apprenticeships in the Trade or Business of a Clothier, and in the several Branches of the Woollen Manufacture, of Weaving, Dying, and Cloth dressing, in the several Counties of *Gloucester*, *Wilts*, and *Somerset*; and likewise for permitting the Exportation of certain Cloths unshorn."

IT may be more convenient to divide the following Observations into such as affect the Bill at large, and such as more immediately affect the very numerous body of Clothworkers or

Shearmen, upon whose behalf they are drawn. The former relate to the regularity of the Bill, according to the usage of Parliament. The latter comprise matter of great public consideration.

This Bill is evidently founded upon the favourite theory of Dr. Adam Smith, that Manufacturers should be under no restraint whatever, and that even the ancient and almost parental system of apprenticeship is inconsistent with the freedom of trade. It therefore proposes to repeal in substance the whole code of laws instituted for the government of the Woollen Manufacture between the 13th of Rich. II. and the 13th of Geo. I. and to dissolve and annihilate the law of apprenticeship.

It may be remembered that when this Bill was first introduced, the Chancellor of the Exchequer expressed his desire, that, as it affected the great staple of the country, and was, on other accounts, of immense importance, such Acts and parts of Acts as were proposed to be repealed, should be particularly set forth, so as to shew their tendency; and that the amendments and regulations meant to be introduced, might also be fully stated, so as to shew their policy. It was answered, that the Bill would speak for itself. Does it do so? On the contrary, it is a mere sweeping Bill of indiscriminate repeal. It does not set out the title of a single Act meant to be repealed. It quotes no part of their full and argumentative preambles or clauses; it amends nothing—all is repeal and annihilation.

The reason why neither the titles or the preambles of the Statutes proposed to be repealed, are set out, is obvious; they are in general remarkably declaratory, not only of the policy of their respective enactments, but of the extreme circumspection, and the deliberate caution with which the Legislature has proceeded with regard to this favourite and national Manufacture, during a progressive experience of between three and four hundred years.—To have contrasted this recorded circumspection with the bold and innovating spirit, which seeks, as it were, by the dash of a pen to obliterate a body of institutes, might have been fatal to the views of the small but affluent and powerful association of Master Clothiers, who seek the completion of their fortunes in an utter freedom from parliamentary restraint.

That those gentlemen have objects infinitely beyond what they avow to Parliament, will become apparent, on comparing the allegations of the Bill, with the enactments which follow: the latter should, and usually do, flow from the former, as a necessary consequence. In the present instance, nothing can well be more loose and general than the allegations; neither are they the evidence, or regular ground-work of the enacting clauses.

The Bill first alledges that the prohibition of the 5th and 6th of Edward VI. respecting the use of Gig Mills, and the Act of 2d and 3d of Philip and Mary, respecting looms, if inforced, would be very prejudicial to the *Woollen Manufacturer*.

2dly. That *many* of the before-mentioned Acts (that is between the 6th of Edward VI. and the 13th of Geo. I.) have been for many years considered obsolete, and that the Woollen Manufacturers and others have in consequence incurred *various* penalties.

3dly. That prosecutions have lately been *threatened* on *many* of the said Acts, and *doubts* are entertained whether *some* of them are not repealed by subsequent Acts.

4thly. That permitting *such* prosecutions to be proceeded in, and the said penalties, &c. to be inforced, would be attended with ruinous consequences, to the said Manufacturer's trade, and to the persons engaged therein.

From these allegations, one would naturally have expected a system of wholesome and salutary regulation, suited to the present period, and always keeping in view the uniform policy of the state, to promote and preserve as much as possible, a high reputation for the *quality* of our Woollen Cloths; some rules for instance as to what kind of Cloths so powerful a Machine as the Gill Mill might be applied to without stretching them to a degree inconsistent with the credit of the fabric, and some regulation as to the number of looms to be kept by each Manufacturer; upon which latter circumstance is known to depend that great question of the domestic or factory system of carrying on the trade. It was also natural to expect that something like discrimination would have been made as to which of those acts between the 6th of Edward VI. and the 13th of George I. are considered as obsolete, and that some specification would have been made as to what penalties had been incurred and under what acts—as little does it appear what prosecutions have been *threatened*, or respecting what acts doubts are supposed to exist as to their being repealed or in force. Nothing of this kind appears; but the bill proceeds to enact, by way of preventing prosecutions respecting *looms* and *apprentices*, that hereafter the Manufacturer shall be under *no restraint whatever* as to the *manufacturing* of Woollen Cloths, or as to the *sale* of such Cloths; and by way of curing *doubts* as to what Statutes may, or may not be repealed, it in fact annuls them all. Searchers, Sealers, and Overseers are to be dismissed; sealing, stamping, and measuring are to be done away, and the Clothier is to be left at liberty to strain his Cloth six yards upon forty, or seven yards upon thirty two, (vide the evidence) beyond the legal stamped length, in order to deliver him from his apprehensions of prosecutions respecting an excessive number of looms, or the employment of illegal workmen; the only prosecutions which have been *threatened*.

As to the other main feature of the Bill, namely, the abrogation of the system of apprenticeship, how far such a measure is wise, virtuous, and just, either in itself, or as it respects the Clothworkers and Shearmen, who have, with very few exceptions, served seven years to the trade, will be considered under their particular case. It is only necessary here to observe, that it now turns out that the witnesses who appeared before the Committee upon the Petition for the Bill spoke (as they have since admitted themselves) under an utter misconception of the law of apprenticeship. It seems, that when they stated the small number of apprenticed journeymen, they meant to confine themselves to those who had actually served their time under *indentures*, their conception being that all others, whether masters or servants, were the subjects of prosecution, notwithstanding they might have worked at the business seven years, which, as has been repeatedly decided, satisfies the statute as completely as if they had been indentured.

The preceding observations are made to shew one irremediable defect in the Bill, namely, that the allegations, even if proved, do not warrant, or shew the necessity of the enactments. The allegations relate at most but to partial inconveniences, the enactments are not those of remedy and regulation, but of the absolute repeal of the substance of one hundred Statutes, which as is observed by an eminent writer, the English Parliament has planted by way of guards and fences round its favourite Manufacture.

If further observation were necessary to shew the serious and extensive effect, of loose and general phraseology in Acts of Parliament, the following fact would be sufficient. The Counsel for the Bill, contended that the word *Woollen Manufacture*, comprised the *Worsted Manufactures*, which, though not of Cloth, are still Woollen; the gentlemen are undoubtedly right in their exposition; but what Hon. Person imagined, when he consented to the introduction of this Bill, that he was involving in its operation, a Manufacture, which has hitherto been groverned by a distinct code of laws, and for the regulating of which a great number of separate Statutes have passed from Richard II. to the present time.

The parts of the Bill, which more immediately affect the Clothworkers or Shearmen (terms which always mean the same person) are the 1st. section which repeals the Statutes of the 5th and 6th of Edward VI. prohibiting the use of Gig Mills; and the third section which repeals the Act of the 5th of Elizabeth, 'enjoining apprenticeship, so far as it affects the Woollen Manufacture in certain Counties.'

Previous to the particular discussion of these

sections, leave is requested to offer some general observations.

It has been ingeniously attempted to confound the opposition to the Gig Mill, with an opposition to Machinery in general; and evidence was even examined as to the disturbances created many years since, by the introduction of Spinning Jennies, Scribbling Machines, &c. &c. The Petitioners beg to be understood as confining their case, and their opposition to the *Gig Mills* the only Machinery mentioned in the Bill; the other Machines alluded to, are now of considerable standing the Petitioners have not even hinted an objection to them. They form no part of the present question*. Much earnestness was also discovered, and several witnesses were examined to shew that riots had taken place, that associations of workmen existed, and that the defence of their interests before Parliament were defrayed by subscription. Can or ought these facts to influence the discussion on a Bill, professedly affecting the staple Manufacture of the kingdom, and which must of necessity involve in its con-

* In a pamphlet lately published by Mr. Anstie, the Petitioners are treated thoughout as persons who are opposing *progressive improvements on Machinery.*

sequences very many thousands of persons. If associations be culpable, the observation equally applies to the other side. The Clothiers, who seek the Bill, admit that they are an association; that they defray the expence of soliciting this Bill by subscription, which subscriptions, it has been part of their Attorney's business to collect. Dr. Smith observes that combinations take place oftner among the masters than the men, and that such are more effectually dangerous. The Legislature adopting this idea for the first time in a late Bill, has extended the penal consequences of combination to both these descriptions of persons.

It has also been argued as if the Petitioners sought the general prohibition of the Gig Mills. Not so; it is the Suitors for the Bill who seek its universal toleration. Had they suggested any qualification, such as confining its use to particular Cloths, whose texture will endure the operation of the Mill, it might have lessened the objections to its establishment, and softened the apprehensions of the Petitioners with respect to a general dismission from their employment; but instead of this, it is roundly asserted by those who seek the Bill, that they will accept of no alternative, no compromise, no suspension; it must be the whole Bill or nothing. It has also been asserted in argument that the Gig Mill is universally used in *Gloucestershire* and in *Yorkshire*, and that it has been so time out of mind; the fact has been proved to be otherwise. The Manufacture of Gloucestershire, consists almost entirely of white Cloths, that is of Cloths which are afterwards dyed in the piece. These are of a strong kind, and to them the Gig Mill has been more or less applied for a considerable number of years; and within about the ten or twelve last years, hand-dressing of *white Cloths* has been almost wholly discontinued. Whether even those cloths have profited much by the abolition of hand-dressing, may be safely referred to the wearers, nothing is more notorious than that the superfine Cloths, though increased in price, have become for several years past less and less durable, and most serious complaints are known to have been made for some time past to the East India Company by their Supercargoes at Canton, respecting the progressive deterioration of superfine Cloth, and the thinness of its texture; still however the Petitioners admit that after the toleration which the Gig Mill has experienced in Gloucester, to put it down as applied in that County, might create a considerable degree of embarrasment and inconvenience. But it will hardly be contended that therefore this Machine shall be allowed in the County of Yorkshire, where there are not at present more than about ten set up, and most of those recently, or in the counties of Wilts and Somerset, where their introduction has been still more recent and limited. Each of these Counties manufacture a different kind of Cloth, namely, what is called by way of contradistinction medley Cloths, (meaning those, which are dyed in the wool) and which Cloths are, on all hands, admitted to be less capable of withstanding the force of the Machine; a Machine that unless managed with extreme skill and caution, as is acknowledged by the evidence on the other side, almost in the very language of the 5th and 6th of Edward VI. overstrains the Cloth, weakens the texture, and injures the ground.

It has also been positively declared, that the trade *cannot go on* without the use of the Gig Mill; and that unless the statute of the 5th and 6th of Edward VI. be repealed this sessions, the trade *must be ruined.* This assertion will appear the more singular, when it is known that seven-eighths of the Master Clothiers in all the clothing counties, except Gloucestershire, are against the Bill, and have either petitioned against it, or signed their disapprobation.

The Petitioners admit, that if what is called state necessity, or national advantage, demanded this measure, they would not be the objects of privilege, but of regret and indemnification. They are willing however to meet the question on its broadest ground, and they respectfully proceed to shew, all the circumstances of the case considered, that the proposed adoption of the Gig Mill, and the dissolution of the system of apprenticeship, are both measures pregnant with public mischief.

With respect to the Gig Mill, the Petitioners are not insensible of the enthusiastic opinions which prevail respecting the application of machinery to manufacture; but the wisdom of that application must depend upon circumstances; the dispensing with manual labour is in itself a great and positive evil; it gives a fatal check to population, and, in the language of political œconomy, deprives the land of a portion of its customers, a nation's best and first consideration. Under other circumstances, machinery is of infinite importance, and deserving of the highest encouragement. Wise and eminent men have in their writings so clearly pointed out in what cases the advantage is paramount to the evil, that it will not be difficult to determine whether the case in question falls under the former or the latter description.

The Petitioners submit the following proposition as one of the results of the arguments alluded to, namely, that the application of Machinery to Manufacture ought not to be encouraged, unless the improvement of the article, or the saving in price, be such as to make amends to the state, for the number of persons thrown out of employ.

They further contend, that neither ought it to be encouraged, when the foreign market is already in our hands, the price unobjected to, and the customer satisfied; for in such case there is no advantage to balance with its depopulating tendency.

Whether the Petitioners have brought their case within these rules of exception, must depend upon the evidence; of this, they are anxious to take the most impartial, though necessarily brief and general review.

The purpose of the Gig Mill is to raise the nap upon the cloths after they have been felted at the Fulling Mill. The next process is shearing them, or cutting down the nap so raised. They are afterwards hung upon the tenters, where they are dried and undergo some further operation. They are finally pressed and folded for sale.

The Petitioners against the Bill have called several Clothworkers, that is men who have worked in the Gig Mills, and who have accurately described its nature and operation. They concur in stating, from considerable and personal experience, that the cloths which are raised (that is the nap of them) by the Gig Mill are now, as in the days of Edw. VI. overstrained. They mention some instances of as much as 7 yards upon 35 and 6 yards upon 40 beyond the legal length; but that in general the Gig Mill does, and must of necessity, from its force, strain 2 yards upon 20; that the white cloths, such as are made in Gloucestershire, will bear much more than the medley cloths, that is, such as are made principally in Wiltshire, Somersetshire, and Yorkshire; that they have known instances of the excessive straining making many bracks in the cloth, which are fine-drawn, and filled with flocks pressed in before sale; but that even 2 yards upon 20, though it may not break, impoverishes the ground That the straining, when *done by hand*, may be little or more, according to choice, that it is strained in fact seldom more than a yard upon 20, much of which goes in again during the subsequent process, but that the straining by the Gig Mill does not go in again; that, with respect to labour, one man and two boys do the same work by the Gig Mill as eighteen men and six boys by hand; that many Clothworkers and Shearmen are out of employ in the counties of Wiltshire, Somersetshire, and Yorkshire, even at this busy period; that, upon an average, there are a great many more than can find employment, and that, in proportion as the Gig Mills shall be further introduced, the proportion mentioned must be turned out of employ; and that in such case, they must seek some other line of life, or become chargeable, as all the other branches of the trade are abundantly filled up, even if they could turn their hands to them; that a great proportion of the Clothworkers, or Shearmen, have been indentured apprentices, and all have been either indentured, or have served seven years to the business (which equally entitles them to carry it on.)

Several very respectable Master Clothiers confirmed every part of this testimony, adding that they adhere to the dressing by hand from the conviction that the other mode impoverishes the Cloth, that they could if they chose, easily have theirs done by the Gig Mill, and that the extra straining would be to them an extra profit, but that they do not think it consistent with their reputation as Clothiers, to pursue that mode, that the number of persons who seek the repeal in question is very few compared to the body of the Master Clothiers in the Counties of Wilts and Somerset, where the superfine medley Cloths are principally made, that are against it, that in the principal clothing towns of these counties such as Trowbridge, Chippenham, Frome, Bradford, Melksham, Beckington, &c. &c. there is a very great majority of the Masters, such as thirteen out of sixteen against this Bill of repeal, that in the opinion of some of them the object of the few persons in question is by means of their large capitals, Mills and Factories, to get the trade into their own hands, and a conversation is stated to that effect. Two of these gentlemen in particular, Mr. Fowle and Mr. Awdry of Chippenham, the former, one of the most eminent Clothiers in that part of the country, have both made the experiment of gig-dressed

Cloths, and abandoned that mode from the injury which they conceived it did to the texture.

Some Packers and Pressers in town, persons having no interest in the question, stated the considerable number of gig-dressed medley Cloths which they found injured, when passing under their observation, and the great complaints made by the London Drapers on that account. —A fine-drawer from one of the factories proved her constant employment in fine-drawing, flocking, and concealing the bracks made by the Gig Mills, and how little of that sort of attention the Cloths dressed by hand comparatively required. A master Dresser from Bradford stated the progressive necessity of discharging his men within the last three years, as the Gig Mills were progressively erected.—Four witnesses proved similar facts as to the County of Yorkshire.

This body of evidence has undoubtedly been met by the testimony of several gentlemen of great affluence and respectability, who assert that the Gig Mill has produced the most beneficial effects, that it renders the Cloth much more acceptable to their customers; that it is finer and softer, that it does not strain more than hand-dressing, that the Machine is perfectly under the controul of the manager, that it has extremely increased their trade, and that without it they could not have executed their orders. They admit the number of Clothworkers that must be thrown out of employ by their more general use, but contend that other employment may be found; and a surveyor of the roads is produced, to prove that there is at present a want of hands to break stones on the road, to wipe off the dirt, and pile them, as well as to dig in the making of canals, and to wheel away the soil; they impute the non-employment of the Shearmen, to their obstinacy in refusing to shear after the Gig Mill, and say, that unless the Gig Mill be allowed to the superfine medley Cloths, the Counties of Wilts and Somerset must be ruined; that their Cloths receive no injury whatever from the Gig Mill; that they even fetch higher prices, and have been generally sold for more than those done by hand; that the price of the raw material, and workmen's wages have advanced 100 per Cent, and it is owing to the Gig Mills, that Cloths at the same time have not advanced more than 30 per Cent; that Mr. Austin, of Gloucester*, his Cloths are all dressed by Gig, and are preferred at foreign markets to all others, and that however great the majority of Master Clothiers may be

Mr. Austin's goods being preferred, is no proof of the superior goodness of *Gig*-dressing to *Hand*-dressing, as none of these goods (white Cloths) are made any where but in Gloucestershire—and is therefore only the preference of one Gig Cloth to another—probably on account of the goodness of the Wool.

in point of number against the Bill of Repeal, that all the respectable part of them are for it, meaning by that, their extent of business, upon which * principle it was asserted, that 7-8ths of the capital and business of the trade are for the Bill.

When witnesses are found in contradiction to each other, the mode generally adopted by impartial and discriminating judges, to get at the real fact, is to look at the situation of the contradicting parties, to see first whether the one may not have superior means of information to the other, and next, whether the particular predicament in which they stand in point of interest, may not be such as to bias the judgment of an otherwise correct mind. A judge in such a case would still more scrupulously examine whether, amidst a great deal of general testimony, certain admissions may not have escaped from witnesses, sufficient to satisfy an enlightened and penetrating mind, as to the true state of the case, it therefore becomes necessary briefly to examine upon these principles, the result of the cross examination of those gentlemen who have appeared in support of the Bill of repeal†.

* This is amply refuted as far as respects Wiltshire and Somersetshire, at least, which were the Counties then in question, by an estimate taken upon the spot since, and in consequence of the above testimony.

† It was intended here to have noticed some particulars of the cross-examination of the Suitors for the Bill; but to avoid the possibility of the suspicion of any partial selection, it is thought better to refer the reader to the evidence, which is printed by order of the House of Commons.

In addition to these particular points of evidence which have been thus adverted to, it may be necessary to notice some of a more general nature. It is observable, that not a single person who has worked in a Gig Mill, and had experimental knowledge of it, has been called. It has been said, that without the Gig Mills, the orders could not have been executed. This however, upon further interrogation, is admitted to mean Machinery at large, such as the Spinning Jenny, &c. introduced many years back. As far as respects the Gig Mill, Mr. Fowle, and several other eminent Clothiers, expressly declare they have never known a greater demand than might easily be executed by hand-dressing.

The gentlemen possessed of Gig Mills have, in strong terms, imputed the increase of their business during the last ten years to the use of that particular machine. This opinion they have themselves uncontrovertibly refuted, by admitting that the Woollen Cloth Trade at large has progressively increased during that period, and that the trade of those who dress by hand has increased in the same proportion. The fact is, that the French Revolution has occasioned the great difference, by casting, as it were, the Continent upon us for supply. It is recorded, that when a much lighter plague than the Revolution, (being a disease only which infected the body) raged at Marseilles, England and Ireland were unequal to the execution of the vast orders which they received

Upon these facts the question is, have *the suitors of the Bill* made out such a case of national necessity as will justify the positive evil of throwing so prodigious a proportion of the Woollen Manufacturers out of employ by the repeal of the Statute which prohibits the Gig Mill. It was admitted by counsel and witnesses on the other side, that the difference from the use of the Gig Mill only, would be as three to twenty-four—The number of Clothworkers and Shearmen in the Counties of Berkshire, Wiltshire, and Somersetshire, with those of Gloucestershire, who are likely to be affected by the application of the Mill to medley Cloth, are supposed to be about eight thousand. According to a late calculation, stated in the House of Lords, from actual enquiry concerning the prisoners for small debts, it appeared, that 1,125 men had 500 wives, and 1,400 children.; to deal in round numbers, say that 1,200 men have two 2000 wives and children; according to this calculation there would be 8000 men, with about 13000 wives and children dependant upon them, thrown out of employ, (with the exception of one eighth part) by this fatal Machine. The evil of dispensing with the labour of man is undoubtedly positive and great. The land is deprived of so many customers, and instead thereof has so many incumbents to support. There are, however, situations in which a country may find it necessary to make the sacrifice; for instance, if our rivals used this Machine, and it improved the quality or lowered the price of Cloth, we must follow them or lose the trade.—If they undersell us, we must resort to any means, however depopulating, of counteracting that advantage, for the alternative is a part or nothing; but when as admitted, and even vaunted on the other side, that we undersell foreigners; that we are in complete, and according to all appearances, in permanent possession (unless we drive our workmen into the already opened arms of our rivals) of the foreign market; and when our customers are satisfied both with the price and quality of the commodity; where is the public inducement for driving so many thousand persons to distress and despair, unsettling the trade, embarrassing and injuring an immense majority of the masters, and dispersing, beyond the possibility of recal, the most valuable body of workmen ever known to this or any other Country.

The principles upon which Machinery may be encouraged are so ably laid down by eminent writers upon commerce and political œconomy, that it is thought more adviseable to refer to them than for the Petitioners themselves to enlarge upon the subject. Machinery is undoubtedly pregnant with public benefit, but like other advantages it has its boundaries, beyond which it becomes a mischievous perversion and national evil.

Extracts from Postlethwayte's Edition of Savary's Commercial Dictionary, Titles, WOOL—CLOTH —*and* MACHINERY.

"THE increase of our Woollen Manufactures will keep up the price of the product of the lands, as demands for goods, not only employ our poor, but cause the advance of the price."

"The products of the land of England is a great part of every Manufacture, our rents are but the value paid for the products of the lands.

therefore all the additional labour we encourage, which pays to the product of the lands, is so much added to the rents of the kingdom."

"An increase of wealth may attend a decrease of numbers and real strength; for which reason the balance of money is less to be regarded, the exports of our nation may, at present, exceed a million in the lower branches of bays, serges, druggets, and flannels, and this may employ a *million* of hands. In a course of years we will suppose this trade to be changed for that of wrought works, tapestry, painting, and statuary, in which our exports might amount to *two millions,* for which *a thousand hands,* full employed, would be more than sufficient."

"Here we see it possible that our wealth may be annually increased *a million,* whilst our *real strength* is decreased in the proportion of *a thousand to one.* There is nothing incompatible in the arts of ingenuity and those of labour; and all the plainer trades may be retained, notwithstanding the higher arts are introduced, in which case we should be both a richer and greater people."

"In such a variety of reasons it is not easy to determine to what degree the use of machines in general should be admitted. States without commerce, regard chiefly the increase of numbers, and their home markets; and commercial states are too apt to consider wealth alone, and foreign markets;—machines without prejudice to either, may be allowed in the following cases:"

"When they do such business as cannot be performed by hand at all; of this kind are pumps, fire engines, water engines, looms, wine, and oil presses, hand-mills for grain, and perhaps horse-mills."

"Where the commodities wrought by them are such as would not be used at all, except they were done by the machine, either being not cheap enough or not good enough for consumption, when prepared by hand. Under this head are the mills for making paper, those for forging and slitting iron and copper, fulling of cloth and leather, and making gunpowder."

"A people without commerce may safely refuse to admit stocking-looms, sawing-mills, throwing-engines, weaving or spinning-engines, mills for striking-files, cutting watch-wheels, making nails, and all the variety of inventions produced by a rivalship amongst nations contending for commerce, and private men for orders."

"Commercial states must have their eyes on their neighbours, and, if they design to engross foreign markets, must provide for the cheapness of labour at home. Goods must be made cheap to render them of general use abroad and at home. Engines for shortening business ought to be rejected or not admitted in commercial states, when the commodity is not all sold abroad; when it affects not the price of labour, *when the machines would lessen our home markets more than increase our foreign ones.*"

Montesquieu says, "those machines which are designed to abridge art, are not always useful. If a piece of workmanship is of a moderate price, such as is equally agreeable to the maker and buyer, those machines which would render the manufactory more simple, or in other words diminish the number of workmen, would be pernicious *".

"M. de Boulainvilliers in his view of the commercial interests of France, enters more particularly into the debate. "A malady," says he,

* Montesquieu, vol. ii. chap. 16. p. 290.

is expanded over our arts. I would speak of machines which tend to simplify and diminish the labour in our manufactures. The cabinets of our ministers are every day filled with projects for machines proper for lessening the number of hands employed in our fabrics. From hence it results, that a multitude of artizans in France are become useless, and must offer their industry to others. I do not speak only of springs and movements; all the manufactures of the kingdom will soon form themselves—they will be in no want of artizans. The first object of manufacture is employment, or what is the same thing, of giving subsistence to a great number of men. Wandering from this principle is contradicting the institution itself of the arts; it is diminishing the number of men; for every time that we place bounds to subsistence we do the same to population; so established is our prejudice, and so generally is it received that we are come to confound the produce of industry with the means of subsistence, which ought to create industry, two things very different in their principle and object. Let us suppose a manufactory to employ 10,000 citizens, and on the other hand 500 machines which produce the same amount of labour in the same art. It is certain that the first gives a living to a much greater number of subjects, and therefore compleats much more important object. The maintenance of 10,000 artizans who draw their subsistence from a branch of industry, forms in the state a multitude of other smaller manufactures. But to set the inconvenience of this prejudice in the stronger light, we must begin by curing another. I speak to those who establish it as a principle, that when a machine diminishes the artizans of one art, that they directly apply themselves to another. Since the taste of mankind is so very refined, and that one great luxury always succeeds another, the arts become so complicated, that a man has not time in the course of his life to master above one. If he loses that by any accident he can never have any hope of replacing it; if it was not so, we should not find, in every revolution which happens in an art, so many idle men, and so great a number of useless subjects. It is objected (and here lies the strength of their system) that machines in diminishing the price of labour bring in great riches to the state; but they do not bring in to their account the precarious manner in which these riches are distributed. They concern only a small number of particular persons, the proprietors of the machines. They cannot demonstrate geometrically that a sum of ten millions, which circulates amongst fifty thousand manufacturers, is better for the state than one of an hundred which circulates amongst a thousand*."

It was more than once insinuated, that there is plenty of employment for the Clothworkers in the other branches of the business; this was contradicted by the most indubitable testimony, and it stands virtually refuted by the suitors for the Bill, who shew what kind of employment may indeed be found for these disbanded artificers, all of whom, with scarce an exception, have served seven years to the trade, confiding in the protection of the existing laws; many of them have now large families, and are bringing forward their children in the same line, and others have reached the latter days of life, with neither strength or spirits left for new undertakings. What then is this employment? Can it be believed? A surveyor of the roads is called gravely to state to the Committee that he wants hands, and that these formerly apprenticed Clothworkers may find employ "in breaking stones, wiping the dirt from them, piling them, and in digging of canals, and wheeling away the soil!" Where the sensibility, where the discretion, where the decency was of calling such a witness, is left for those

* Les Interets de la France mal entendu, Tom. 3. p. 272.

Clothiers to explain who brought him forward to give his testimony.

It is highly observable that two of the three propositions with which these observations set out, are admitted; namely, that the legal establishment of the Gig Mills would throw out of employ an immense body of Clothworkers or Shearmen; and that these Machines have not hitherto, during their long use in Gloucestershire, or their partial and recent use in the other Counties, lowered the price of the commodity to the consumer; on the contrary, every witness has admitted that the Gig-dressed Cloths have been sold as high, or higher, than the hand-dressed Cloths. They admit also, that they do not cost so much. One gentleman indeed has said, that, considering the advance of the raw material, and of workmen's wages, nothing but the Gig Mill could have prevented the Cloths from being raised no more than thirty per cent. This assertion likewise the Clothiers themselves refute upon their cross-examination; for if the Gig Mill has prevented a higher rise, those who do not use the Gig Mill must have been compelled to advance their Cloths higher than those Manufacturers who employ that machine; instead of which, themselves admit that the hand-dressed Cloths are sold somewhat lower.

As to the third proposition, namely, the over-straining, and ultimate deterioration of the Cloth, after admitting that such is the fact with respect to the public Gig Mill, which is the description of much the greater part; it is said that there is one most complete and satisfactory security against the impoverishment of the Cloth, namely, *the interest of the Manufacturer*, who would lose his trade if he sold a bad commodity. Upon this principle and mode of reasoning there is scarce a Statute upon the books, that would not be superfluous, does it not apply in particular to every law, respecting trade and Manufactures? But may not a man's *immediate* interest be so great, as to render his *remote* interest a matter of indifference? And is not this the case in the present instance, and does not the evidence prove it?— Mr. J. for instance, is stated by his Clerk, to make thirty and upwards, Gig-dressed Cloths per week. It is admitted that three persons will dress four Cloths per week, by the Gig Mill, which, by hand, would require twenty-four —suppose them on either side to be men—say,

4 Cloths, 3 persons, at							
2*s.* 6*d.* per day -	0	7	6				
A week of six days -			6				
	2	5	0				
1 Piece ¼	0	11	3	per week.			
Say, 30 pieces, at 11*s.* 3*d.* is per week,					£16	17	6
4 Cloths, 24 persons, at							
2*s.* 6*d.* per day	3	0	0				
			6				
To make the same a week of 6 days,	18	0	0				
One Piece ¼ - -	4	10	0	perPiece			
Say, 30 pieces per week, at	4	10	3	is	135	0	0
				Difference, per week,	118	2	6
				Difference, per year,	£6142	10	0

From this statement, it appears that the Manufacturer dressing by Gig Mill saves in wages only £6412. 10s.

It is apparent, from the tenor of the evidence, construing it in the most impartial manner, that a further and very considerable profit results from the straining the Cloths beyond what is called the legal measure, or one yard upon twenty. Several witnesses, who have had personal experience *as workmen*, declare that they have frequently known the straining to be as much as 6 yards upon 40, and in one instance 7 yards upon 35; that they have never known less than 2 yards upon 20; and they do not think the Gig Mill can be so controlled as to strain less; that they have always understood it to be the wish of the masters to strain as much as they can, without breaking the ground. It is admitted, on the other side, that Cloth might be strained 2 yards upon 20, without breaking the ground, though the Cloth would be lighter, and weaker in its texture; in return for which inconvenience the Gig Cloth has a finer gloss, and by extra straining feels softer to the hand; that some persons may perhaps have their Cloths strained more than the witnesses have, particularly those who have them done at the public mills. But, at all events, whatever the straining may be above 1 yard in 20, it is admitted to be the profit of the Manufacturer, for whom the Cloths are thus dressed. Under these circumstances, it cannot be thought an unreasonable statement to suppose that, on an average, an extension of 2 yards upon 20, that is one above the legal length, either does now or will take place, if the Gigs shall become legalized. And let us still suppose the medium of trade to be like Mr. Jones's, 30 pieces per week.

The pieces of Cloth average about 30 yards in length, say,

Upon a piece of 30 yards extra straining, 1½ yard, at 17s. per yard — — — —	£1	5	6
30 pieces per week, at 25s. 6d. — —	38	5	0
Per annum	£1989	0	0
Result. Saving by wages — — — —	6142	10	0
By extra straining — — —	1989	0	0
	£8131	10	0

Here is an extra profit to the Manufacturer of upwards of £8000 per annum. If interest be the ruling principles of the trader, surely all these gentlemen who have given testimony, have the most powerful inducements; long before the reputation of the fabric can become so affected as to turn the channel of the trade, especially under the present circumstances of Europe, which must for a long time, as it were upon compulsion be our customers, these few gentlemen will have amassed princely fortunes! The impoverishment of the Cloth may be progressive, but the legislature will look to consequences; every person must be conscious that superfine Cloth has for several years past become less and less durable. It appears by the evidence that Spain and Portugal are calling for stouter

Cloths, and complaints of the most serious kind are known to have arrived from the East India Company's super-cargoes at Canton, repeating and referring to similar preceding complaints, that the texture of the Cloths become weaker and weaker, so as to threaten the continuation of this immense branch of export.

Upon these grounds the Clothworkers or Shearmen most humbly submit, that the calamity impending over themselves and families is not called for by national necessity (and which if made out would leave them as objects for in demnification not destruction) but that on the contrary the price of the commodity is not lowered to the consumer or foreign purchaser by the Machine in question, and that its tendency is the manifest and certain, though perhaps progressive impoverishment of the quality and reputation of the national fabric of Woollen Cloth.

The other part of the Bill, which affects the Petitioners. namely, the proposed repeal of the 5th of Elizabeth, respecting apprentices; and which, as to the Counties mentioned in the Bill, entirely dissolves the ancient system of apprenticeship, however deeply it would affect the Clothworkers or Shearmen, if passed into a law, does not call for any lengthened animadversion. Considering the antiquity and universality of that system, considering that it was not originally founded upon mere trading, but upon great moral and political principles; considering that its annihilation, would inevitably shock and convulse the whole Manufacturing interests of the Country, and considering how ill the millions of persons employed in them, are prepared for this great commercial revolution, which, if it takes place at all, must be universal. One is led to enquire what is the ground? What the alledged imperative necessity? What the immediate and impending danger, which calls for this perilous innovation? It is not easy to suppress one's astonishment at finding the single reason, alledged to be *threats* of prosecution, against non-apprenticed masters, and non-apprenticed journeymen! That astonishment is increased, at finding even this shallow pretence for so awful a change of system almost entirely abandoned before the Committee. The gentlemen admit that they formerly gave their evidence under an impression that all were carrying on the trade illegally who had not been absolutely bound by *indentures*, but being now informed that all who have *exercised* a trade for seven years, are entitled by law to carry it on, they can state but very few instances indeed; out of the very many thousands engaged in the Cloth-working or Shearing trade, who have not exercised it seven years!—but supposing their assertions before the Committee, which set on their *Petition for the Bill* to have been sustained, namely, that if the law were enforced against illegal workmen, there would be none left to work at the trade, and to that add the consideration of what certainly is the law, namely, that the Masters are liable to prosecution for the employ of those young persons in the trades mentioned in the Act; such as weaving, dying, and, shearing, who are now in the course of their customary *parole* apprenticeship—how simple is the remedy! *indenture* these young persons for the remainder of their time, and apply to Parliament for an Act to indemnify those masters who have unwittingly offended the law. It is not pretended that prosecutions have been even hinted at, as to this latter part of the case, parole apprenticeships being the universal custom of these counties, and approved of both by masters and men. But why not *indenture* these youths, let the revenue derive its just advantages, and thereby encourage both the youth and his master to cultivate that moral and almost kindred feeling towards each other, which distinguishes those trades and those places, where the custom of indentured apprenticeship prevails.

Much pains have been taken on the one side, to shew that the trades in question may be learned in a few months as well as in seven years; while, on the other side, old and experienced masters, who have trained many apprentices, insist that, all circumstances considered, respecting the general tempers and capacities of youths, and the various details of their trade, that seven years are necessary to make a complete and steady workman. Be this as it may, suppose six years, or even five be sufficient, this is matter of regulation, which the wisdom of the Legislature can determine, but surely not a cause for the annihilation of a system! If the Master Clothiers (who are suitors for this Bill) mean, as they say, to insist upon the latter, it must be from motives very different from those which they avow. One of those motives is believed to be the system of factories—a system, which has been found too replete with shocking consequences of every description to be encouraged, except when the interests of trade imperatively require it. Most of the gentlemen who have appeared as soliciting this Bill, and who have erected Gig Mills, have also factories, this being a system which the capitalist only can embrace, and which produces to him a virtual monopoly of the trade. To make this system completely advantageous to him, he feels that all restraint must be done away; the moral obligations of the covenanted master become irksome in these pleasurable days, and the privilege of dismissing young creatures of either sex the moment that disease or other incapacity shall discover itself, (although the effects of their congregation) is too advantageous, in a pecuniary point of view, not to be desired by those who do not hold themselves called upon to look to any other consideration *.

* The nature and tendency of factories cannot be better or more authentically ascertained, than from a late Statute of 42. Geo. III. c. 73. Every Legislative Act records the fact of the evil which it means to remove. It is impossible to peruse the provisions of this, which prohibits the compelling children to work more than fifteen hours in the day, including meal time;—or to work through the dead of night, winter and summer—which provides for their spiritual instruction, and their attendance on divine service; and which forbids males and females being lodged in the same apartments, without seeing what factories were, previously to this benevolent act, had not the facts which induced it been notorious through the country.

In a late ingenious treatise upon agriculture are the following observations. "The influence of this (domestic labour) is very remarkable in the difference which in these circumstances takes place, between Cotton Manufacturers, who in general are collected together in great numbers, often to the extent of more than a thousand; and those who carry on the manufacture of our *National Broad Cloth*. These last almost every where work

It may be necessary here to say a few words upon the law of apprenticeship. This system was not, as has been insinuated, first instituted by the Act of the 5th of Eliz. On the contrary,

separately and unconnected with each other, and employ all the hours which they devote to relaxation and amusement in the care of their gardens, and other small portions of ground that they possess; to which they and their families become commonly so much attached, that they have been known to remain in them, small as their possessions commonly are, for many generations. Nor does this variety of occupation make them worse tradesmen, or induce them to work less, as some speculative men have supposed it would do; while in various ways it tends to make them more virtuous citizens. Their manufacturers are every where noted for their industry, and for the quantity of cloth which they produce: and we all know that the article itself, which has long been considered as the *staple commodity of our country, is the best of its kind that any where can be met with.* Some manufacturers necessarily require the united labours of many workmen; but this is not frequent: nor do we know of any that absolutely require such numbers to be employed in one body, as to prevent the possibility of their being all supplied with small portions of ground. It may be the interest of a few large money proprietors, to collect manufacturers in great numbers together, and thus to enjoy the profits of their accumulated labour; but this being highly detrimental both to the morals and health of those whom they employ, and being thereby injurious to the whole nation, it is surely full time, that to the very extensive length to which undertakings of this kind are frequently carried, some check or regulation should be applied." *Bell's Essays on Agriculture.*

it will appear, from a reference to that Statute, that it recognises and regulates the ancient law. The Act of the 4th and 5th of Philip and Mary, C. 5. (not noticed in the Bill of Repeal), after setting out the evils which want of skill has brought upon the woollen manufacture, positively enjoins that persons shall not in future be put on to work as journeymen, unless they shall have been apprenticed to the occupation, or *have been exercised therein by the space of seven years.* The vast body of workmen in question have been apprenticed to, or exercised in the trade seven years. With what justice or humanity can it now be enacted, as proposed by the Bill, that *all persons* shall be *encouraged* to work at the said trade, although they have never been apprenticed thereto, or worked thereat;* and although those Statutes continue to exist, which impose most heavy pains and penalties upon such artificers as shall emigrate, and after notice refuse to return. In short, they would be superceded and

* An admission to the privileges of apprenticeship, without their having been earned by seven years of servitude, has been bestowed upon the army and navy as a highly beneficial and flattering distinction; and the same is still held out as an honourable stimulus to future exertions—how does this consist with thus opening of trades to all mankind without any condition whatever.

driven from employ on the one hand, and not suffered to accept of engagements in the way of their trade on the other, however liberal or advantageous.

It is now acknowledged law, that either master or man, having exercised the trade seven years, is thereby entitled to carry it on, or to work at it, as securely as though he had been indentured. There are several decisions to this purpose; but perhaps the following extracts from Mr. Justice Blackstone's Commentaries will be thought sufficient to establish that proposition.

"Persons serving seven years as apprentices to any trade, have an exclusive right to exercise that trade in any part of England. This law, with regard to the exclusive part of it, has by turns been looked upon as a hard law, or as a beneficial one, according to the prevailing humour of the times, which has occasioned a great variety of resolutions, in the courts of law, concerning it, and attempts have been frequently made for its repeal, though hitherto without success."

"At common law every man might use what trade he pleased; but this restrains that liberty to such as have served apprenticeship: the adversaries to which provision say, that all restrictions (which tend to introduce monopolies) are pernicious to trade. The advocates for it allege that unskilfulness in trade is equally detrimental to the public as monopolies. This reason indeed only extends to such trades, in exercise whereof skill is required: but another of their arguments goes much further, viz. that apprenticeships are useful to the commonwealth, by employing of youth, and learning them to be early industrious; but that no one would be induced to undergo a seven years apprenticeship, if others, though equally skilful, were allowed the same advantages, without having undergone the same discipline; in this there seems to be much reason. However, the resolutions of the courts have, in general, rather confined than extended the restriction. No trades are held to be within the Statute, but such as were in being at the making of it; for trading in a country village apprenticeships are not requisite, and following the trade seven years without any effectual prosecution (either *as master or servant*) is sufficient without an actual apprenticeship?"

Blackst. Com. Vol. I. p. 428.

If then so few persons can be disturbed by the enforcement of the law of apprenticeship, were such enforcement intended, and if the difficulty respecting those who are liable to it may be so easily removed, it reduces the question of apprenticeship to the mere abstract consideration of its policy; and upon that, the Petitioners do not think it necessary further to reason; they only desire leave to express their humble hopes, that, if an experiment is to be made, that it will not be done at the expence of the sorrow and suffering of so many thousand families, brought up to a trade, the staple of the country, which neither feels nor fears rivalship; which has long been in complete possession of the foreign market, and which boasts the high prosperity to which it has attained, certainly under a system of parole, if not indentured apprenticeship*.

* The following extracts from Savary's Commercial Dictionary (Postlethwayte's edition) may not be unacceptable.

" The commerce of a nation depends on the excellency of its manufactures, and those again on the skill, ingenuity, and dexteriy of its manufacturers, and artificers of every kind. On these considerations we may presume that the French have made such laws to oblige their artists to make themselves thorough masters of their respective trades, and to exclude them from the freedom of the several corporations of arts and trades unless they are capable of performing, with their own hands, what they call a masterpiece of workmanship, that is, such a piece of workmanship as will shew a young man master of his

The Clothworkers or Shearmen of the counties of Wilts, Somersetshire, Yorkshire, and Glouce-

business; and this must not be done in a corner, but in the presence of those sworn for that purpose."

" The national benefit of this policy will further appear from considering the disadvantage Great Britain must inevitably sustain in her commerce, were her manufactures to degenerate in their ingenuity and dexterity."

" However slight some people may make of these considerations upon apprentices, yet the breeding up our young people properly to trade, in every branch of it seems, a matter of the last consequence to a trading nation, and may one time or other be found worthy some people thinking of, when they are once convinced of this important truth, that traders alone are the source of all their treasures, and consequently of all their splendor and magnificence."

" Since such exquisite skill in the manual operation is required by all any way practically concerned in the Woollen Manufacture, do not all our national advantages arise from the Woollen Manufactures, who are to act in every part hereof; for they are not bred in *a few years.* Ought not the wisdom of the nation to be alarmed at the daily artifices, which are used to decoy and instigate our artists and Manufacturers out of the kingdom, to the emolument of other countries, and the certain ruin of this? And can any thing effectually do this, but giving them all due and reasonable encouragement ?"

It is true, that Dr. Adam Smith objects to the system of apprenticeship, as being less emulative, than when youth are left tershire, now submit, with great humility, that they have made out, both by evidence and ar-

to their free, uncontrouled, and ambitious exertions. His observations on this head are incidental to his more general objections to corporations at large; but it is observable, that he considers both as mere questions of trade. He does not seem to have weighed the beneficial political tendency of corporations, so recently and powerfully explained in the House of Commons, nor the moral consequences of the system of apprenticeship. If even parental authority, where command is almost irresistable from the endearments which accompany it, is found too weak to oppose the tempestuous passions of youth, who shall shield them from themselves, when left to the cold unfeeling authority of a mere mercenary task master, whose end is answered, and whose duty is discharged, when the quantum of labour is atchieved. There is not in the world a more virtuous body of men than the Citizens of London. They owe it to their system of apprenticeship. According to the custom of that renowned city, time out of mind, its Citizens, with very few exceptions in certain cases, must have served an apprenticeship of seven years; so long a residence in the family of the master, subject to proper restraint as to the hours of absence or relaxation, and with the example continually before him, of decent deportment, attention to the sabbath, and perhaps family devotion (which is more general in the city of London than levity may imagine), begets habits of thinking and acting, which ultimately render such persons invaluable to their country. A virtuous population is a blessing to a country. A vicious population is but an expanded and contagious evil. If it be necessary to encourage youth by progressive wages, this may equally be done under the system of apprenticeship. Indeed, it is at present universally the custom in the clothing counties.——EDITOR.

gument, the case stated in their Petition to Parliament. The number of them which must be thrown out of employ by the general revival of the Gig Mill has been fully admitted on the other side. They submit, that the only legitimate ground for the public infliction of private injury, is that of the public welfare. They submit, that it will appear throughout the evidence, which is printed by order of the House of Commons, not only from the most respectable testimony adduced by themselves, but from the admissions, either positive or indirect, of the Gentlemen Clothiers who have been examined on the other side, that the price of the article has not been lowered, where these Gig Mills have been used, to the consumer or to the foreign purchaser; and that, although the appearance of the cloth may be made more showy by their use, and rendered softer by extra straining, yet it nevertheless dete-

* It is said, that the Shearmen owe their being out of employ to their obstinacy in refusing to work after the Gig Mill, although it has been proved that disputes have arisen from attempts to alter their wages, and from other causes. It seems to have been overlooked that by the Stat. 39 and 40 Geo. III. cap. 106. sect. 15. the Magistrates may licence unqualified workmen when those who are qualified refuse to work from improper motives.

riorates, and radically impoverishes the texture of the cloth, as complained of by the 5th and 6th of Edw. VI, which prohibits the use of these machines; that these observations apply, in a great degree, to the White Cloths of Gloucestershire, and that all agree that the Medley Cloths of Wiltshire, Somersetshire, and Yorkshire, are less capable of enduring the force of the mill, unless it be very skilfully and cautiously applied, and that the out-dressed Cloths have been much injured thereby.

They submit, that the alleged necessity for the abrogation of the law of the apprenticeship has been in a great degree done away by a proper explanation of the law as it now stands, and a reference to decided cases; and that the remaining inconvenience, which respects only those who are in the course of a parole apprenticeship, might easily be removed, without trampling upon the hard-earned legal privileges of from forty to fifty thousand persons, and bringing down upon them inconceivable suffering and distress.

Under these circumstances, they cast themselves, their wives, and their children, upon the humanity and the justice of the Imperial Parliament.

Printed by C. Stower, Charles-street, Hatton Garden.

OBSERVATIONS

ON

WOOLLEN MACHINERY.

PRINTED BY EDWARD BAINES.

1803.

OBSERVATIONS,

&c. &c.

A TRADE or manufacture is deemed valuable to a country in proportion to the number of hands it employs, and the families it maintains: because in them, is national strength, and the produce of their labour, is the truest riches. The solidity of this maxim is strikingly evident, when applied to the preparation of any articles destined for the consumption of foreigners. Since we can instantly perceive, that the more of our families which can be supported *at their cost*, the more we increase our population and the more able we are to defend ourselves against them in cases of necessity, and the greater wealth we draw from them into our own coúntry. On this sound principle it is, that the exportation of our wool unwrought up is strictly prohibited. But we seem now in a great measure to be losing sight of the leading object of this wise regulation of our laws, which was undoubtedly not alone to add to the value of our wool by labour upon it prior to its leaving us; but chiefly to cause the greatest number possible to derive advantage or subsistence from this prizeable home production—this boon of Providence to our isle.

But

But now on the contrary, the prevailing solicitude seems to be, to confine the advantages presented to us by so valuable a possession, to the fewest number possible, and ingenuity is incessantly on the stretch to accomplish this; and to convert our wool into cloth, as far as ever practicable, by mechanical contrivances, without the intervention of human labour; or in other words, to reduce the great staple trade of the country, to something very little better than the mere exportation of the raw material.

The pretext held out, for this great deviation from the wisdom of our ancestors, is improvement, and extension of trade. *But it cannot be extended beyond what the fleeces of our sheep will permit.* And as there is no complaint whatever, that any wool remains on hand unmanufactured, but the very reverse; it is manifest that the desire for more machinery in the woollen manufactory proceeds from, in fact, not public, but private advantage. It is wholly a race amongst individuals. Public good is certainly out of the question. It is in reality, each one striving against the rest, by every possible means, to draw to himself a large proportion of the business. The means chiefly employed to accomplish this, are mechanical contrivances; every one endeavouring to carry them farther than another for his own peculiar advantage. A line of competition, which has already done much public injury

injury in the woollen trade, and will yet do much more, if not arrested and restrained by the regulating power and wisdom of the Legislature.

However powerful the machinery made use of, no more cloth could be produced—no more could be sold than is already sold; if all the wool that is to be had, is now made up, of which there can be no shadow of doubt, from its continual advance. Neither would there be any wisdom in such a case, to reduce the price of cloth by means of further machinery, (if it would have that effect, which however as we shall see by and by is not quite so certain) as it is very plain, that if we can sell all that is to sell, on an average suppose of five shillings per yard, it would be the extreme of commercial folly, to weaken the population of our country, and lessen so considerably its wealth, by endeavouring through machinery, to let the foreigners have, at 4s. 6d. what they would otherwise chearfully give us 5s. for; unless we could sell them so much more of the article to compensate for the difference and the loss; but that in the woollen trade is totally impossible. To give an instance in conformity to this idea: Suppose the full and complete adoption of machinery for the shearing and finishing of cloth, what would be the consequences? The coarser kinds, which is the bulk, might be sold probably 1d. or 2d. a yard cheaper, the finer ones from 4d. to 6d. lower, but equally imper-

imperceptible in the whole price; not a yard more would or could be sold on the whole, than is already: but we should surrender to our purchasers the said pennies and twopences, that is, the maintenance of thousands of our families—throw them on the parishes; and after thus very foolishly giving up, what sufficed for their support to no manner of purpose, we should have them to maintain out of our own pockets instead of those of the foreigners, in the shape of poor assessments: to say nothing of the mortifying poignant feelings of the men, on being turned adrift from a business, which after a long learning procured them at last a comfortable subsistence, to a poor-house, or some precarious employment if any other could be found, of which they knew nothing.

Hence it is plain that what might be very great improvements in some branches of trade, are not so in the woollen; because it cannot be increased beyond the quantity of wool grown. It is also plain for the same reason that machinery ought to be applied to wool with great caution, and only when absolutely necessary. While on the contrary, some other of our important manufactures, such as the cotton and flaxen, could not on account of the high price of our hand labour be retained, or perhaps exist at all in the country, without the utmost use of machinery.

It is also further evident from hence, that the highest state of perfection of the woollen trade—the way to make it the most profitable to the country, and to let the greatest number of its inhabitants participate of its benefits, would be, to work up the wool as much as possible by hand labour, provided the whole that can be grown could be so worked up in due time; and the goods after such expensive hand labour could meet with a ready sale. But such a nice arrangement and exact proportion of the produce of our sheep to manual labour, and to the prices of woollens in foreign markets, is not to be expected. The farther however that circumstances compel us to deviate from such a desirable state of perfection, and the less advantageous the whole of the community considered, the trade most certainly becomes. But more or less we must deviate, because, by endeavouring to employ the greatest number of hands, and confining the manufacture chiefly to manual labour, we might very much injure the farmer, by leaving him burthened with wool, at the same time that he is forbid to dispose of it elsewhere; and by the high price of hand labour in this country, in comparison to that of others, the merchant might also be beat out of foreign markets by their own manufactories, which now almost every where abound, notwithstanding the comparative superiority and cheapness of our wool.

To prevent or rectify this order of things, so prejudicial to the farmer and merchant, machinery may be advantageously called in as an auxiliary to human labour, but clearly only so far as to clear the farmers of their stocks at a fair price, and enable the merchants to maintain their ground with ease in foreign markets. If farther applied, instead of an useful auxiliary, it becomes an enemy both to labourers, manufacturers and merchants, and friendly only to the farmers, because the quantity of wool we know to be of necessity strictly limited, and cannot otherwise than very slowly, be increased. But we may easily conceive so much of machinery to be used, as would work up the whole of the year's growth, for example, in the space of six or eight months instead of twelve.

The amazing clamour that would be caused by this state of things, among the proprietors of mills for wool to feed their works, will readily present itself to every understanding; and that an enormous advance of the article (of which the late inadequate supply of corn for the year's consumption may give some idea) far more than overbalancing the cheapening power of their machinery, must be the inevitable consequence. It is therefore abundantly clear, from this by no means absurd supposition, that *too much machinery* may be applied to wool; and that there is a point when it ceases to be useful, and where besides its exclusion

exclusion of hand labour, it begins to defeat the very end and purpose for which it was adopted —to produce dearness instead of cheapness. To ascertain then this point is the grand desideratum. From the uniformly rising price of wool ever since the free use of machinery, till its present unprecedented height, and the little expectation yet of its fall, may it not be strongly suspected that we have already passed it? And that, if in attempting to remedy, or as it may illusively seem, to keep pace with the evil, we have still further recourse to more machinery, is it not like adding fuel to the fire we wish to extinguish, the very way to aggravate the grievance? Most certainly it is, if wool be already, of which there seems ample proof, too rapidly wrought up. Go on in the error, increase machinery, and we may soon raise wool and woollen goods so high, or impoverish the quality so much, as totally to exclude ourselves from foreign markets; unless our ingenious machinery advocates, can also invent machinery, to create or produce wool in proportion to their multiplication of compendious means for working it up.

But the misfortune is, that a large new mill can be much sooner erected, than additional sheep reared to keep it in motion. Indeed, without some restriction to the slaughter of lambs, or general inclosure bill,, we may pretty safely conclude, that little or no addition can be made to our growth of wool—for the high price both of carcase and fleece during so long a time, can scarcely have failed to urge every breeder, e'er this, to stock to the utmost that his land will bear. —This mode of reasoning with regard to wool, seems to be founded on indisputable truth. But will not at all hold good, as before hinted, when applied to cotton; for this simple reason, already so often noticed, that the one material is limited in quantity, while the other is continually suiting itself to the demand, and may be produced *ad infinitum*. It is therefore scarcely possible to carry cotton machinery too far, as there is nothing to prevent its extension, but scantiness of demand for the manufactured goods. Besides, as all other countries have nearly equal access with us to the raw material, without very superior machinery, dexterity, taste, &c. we should soon have little else to furnish with cotton goods, but our own country and its dependencies. *Whereas our principal advantage with respect to wool, does not result from machinery, but from our sole possession of the article, from its being also preferable for most purposes, and in general cheaper than in any other known.* Machinery as we have seen to a certain extent, may aid, but when in excess, by consuming it faster than it can be produced, will destroy that advantage, by enhancing the price to such an intolerable degree, as

to

to endanger the loss of a great part of the foreign trade, and render desirable some regulating law, to restrain its use and injurious influence, according to the price of wool, after the manner of opening and shutting our ports to corn, as it rises or falls.

But who shall enjoy the privilege of that quantum of machinery which might be deemed salutary, is a question which at once presents an insuperable difficulty to the existence of such a law. There is, therefore, perhaps no other alternative, if people will not endeavour to understand the subject, and see the necessity of abstaining from further machinery of their own accord, than patiently to wait till one extreme as usual in commercial concerns slowly corrects another, and to submit to the losses and inconveniences of the interim.—This interim of suffering, however, will be of much longer duration now, than formerly. Because when once arrived at the extreme of dearness—when the prices can no longer any where be borne, many mills must stop—many manufacturers be ruined—many merchants at a stand for want of vent, before the present riches of the farmers, and the large capitals now employed in the wool trade, can be conquered, and made to give way, so as sufficiently to relieve the manufacturing branches from such a distressing situation.—We seem now on the high road to this situation, and as wool cannot be much increased, there appears to be no means whatever of preventing our arrival there, but by checking the use of machinery. Already several of our articles are either too bad, or too high, for many of the markets abroad; it will soon be the same with others; and how can it be avoided, but by a more provident use of machinery—a slower consumption of wool?

At the first, and for some time after the introduction of spinning jennies, pulling out and twisting from 40 to 60 threads at once in the place of one, and of carding, slubbing and scribbling billies, performing with one man the work of twenty, &c. all seemed in the woollen trade to go on well; and instead of men being thrown idle as they apprehended, webs were prepared so much more quickly than before, that they all found themselves called upon to the looms, and the women and children only were left without work in their own houses. The trade in general consequently flourished. Goods were rapidly and cheaply made up, and quickly sold. The murmuring about the new machinery in a great measure subsided, and in its place an idea begun to prevail, that the more machinery, and the more need of human labour; and this conclusion is probably perfectly right, with regard to cotton or any other raw material that can be produced at pleasure, furnishing so much more of one sort of employment, if it deprives of another, but fails com-

completely with respect to wool. Though the error might not be easily detected, while wool continued in plenty to feed the devouring machines, working away often day and night, and till the old long accumulating stocks were expended, and the calls became too quick and large for the sheep to satisfy. But now the disproportion between the demand and the supply, between the consumers and producers, begins to be felt, and to convince us, that the effect of a too liberal indiscreet use of woollen machinery, is unavoidably advancement of price.

There is also another evil of no small magnitude resulting from machinery, which deserves the particular notice of the middling manufacturers. This is, that during a brisk demand for goods, they are apt to load themselves with machinery, (forgetting that greediness generally overshoots her mark) which will surely prove a snare to them, in the languid time which always ensues. For it is a settled maxim, that their mills must be kept going, to avoid the palpable, certain loss of their standing still. They therefore go manufacturing on, push their means and their credit to the utmost, as well as encourage their neighbours who have no mills to do the same, till an overwhelming stock of goods is accumulated, and finding themselves unable to proceed, must either sell to great loss; give ear to dubious chaps, and

probably

probably lose the better half; or enter into hazardous adventures, which can scarcely fail to be ruinous, in as much, as they must nine times in ten, be made with unsuitable assortments of goods, and under the direction of total inexperience, bringing destruction on themselves, and injuring materially the regular trader.

Such are the bitter fruits of too much wool manufacturing machinery, when in the hands of middling men, very high prices created by the quick calls for, and rapid advancement of wool, or glutted markets, with costly made up, but lowering articles.

Formerly the mode of making cloth in this large cloathing district was as follows:—A class of men with tolerable capitals called woolstaplers, rode over the country about cliptime, to buy up the wool from the growers. They then have the fleece carefully broke into its various qualities, and afterwards sell it out in small quantities thus assorted, to innumerable master manufacturers of little or no capitals, spread around in the adjacent villages. Those master-makers superintend all the remaining operations; have many performed in their own houses, and hire out the rest, to their neighbouring families: the whole of which, husbands, wives, and children, were employed together in their own dwellings, some in weaving, others scribbling, carding, or spinning.

Since

Since the introduction of machinery a new class of men, as machinery, or mill-owners, are concerned; and many of the master-makers have their own mills. The effeet of which is, that wool is very much faster made into cloth, considerably more weavers are employed, and no home work left for women and children. Now after the wool is dyed by the master, it is sent to a mill, where, with the help of a man or two and a few children, it is most expeditiously scribbled, willied, slubbed, spun and made ready for the loom.

Those mills are costly establishments, and must, if possible, be kept in motion: hence, during the last ten years, wool has been consumed so fast, that it has been steadily on the advance, notwithstanding a war in the mean time, more expressly levelled at our trade, and manufactures in particular, than any preceding one; and at some periods of which severe losses were sustained by the merchants, and almost total stagnations of sale ensued.

But so much machinery being introduced, the manufacturing at all events went on. Wool therefore continued in demand, and on the rise, whatever became of the pieces manufactured; and we have often in the course of said ten years, been presented with (as the fruit of so much machinery) the distressing spectacle, of lowering goods, and rising raw materials, at one and the same instant of time. Wide spread ruin amongst the

poorer

poorer manufacturers was, of course, the consequence—the jails were filled—the poor rates increased—and many, of rather more spirit and enterprize than the rest, unwilling tamely to yield, swarmed over Germany and America with their goods, but still came to ruin. And even now, since the peace, when the usual countries and consumers of our woollens are restored to our trade, at least, so far as French enmity and influence will permit, the price of wool has got so enormously high, many sorts being advanced since the adoption of machinery, not in a double or treble, but even in a four and five-fold proportion, that in spite of every art of the clothiers, the article produced is so bad and dear, that a very large proportion, to be sold at all, must be sold to loss. On which account, and the rapid mode of manufacturing, continual consignments as adventures must be made, and employment furnished to the auctioneers of London, Hamburg, Amsterdam, America, &c.

Coarse and middling cloth-making is undeniably become, a most miserable occupation. No men living perhaps, toil and struggle so hard, and reap so little benefit as the clothiers of the present day. But they know no other business, nor have they capitals to turn to any other; they must therefore struggle on. Wanting the aid of their sons as early as their strength will allow,

they

they are all very generally also, doomed to the same fate, and must become cloth-makers or weavers in their turn.

In the stuff trade this lamentable contest between the oppressive dearness of the raw materials, and the low market prices of manufactured goods, is nearly at an end. Many makers have already sunk under it, and cotton has fortunately furnished a refuge to others. But they have fled to it with trembling steps, when only at the last extremity, and with the chilling apprehension, that one of those stagnations so customary to that fluctuating trade, may soon bereave them of its protection.

It is a curious fact in the said stuff business, that in proportion as mills have increased for the spinning of worsted yarn, which was thought a wonderful improvement at the first; the price of yarn has advanced, instead of lowered, and the number of stuff pieces manufactured, diminished, till the trade is now nearly lost, the foreign part almost totally, not a piece being now made for ten or perhaps twenty or more of former times with hand spinning. The mill spinning however goes on, the profits of spinning being, as it is said, enormous. What becomes of the yarn is not so apparent. Certainly it is not made into stuffs hereabouts as formerly before the existence of mills.

In the cloth trade we are now getting a step further from the old mode of manufacturing above described. Some merchants, in order to concentrate, or rather to avoid all the profits of middle-men, such as clothiers, woolstaplers, master-dyers, salters, oilmen, mill-owners, &c. have themselves become cloth-makers, erected large factories for the purpose, and import or purchase at the source every requisite material. Their considerable purchases of wool, made at once, and that often unskilfully, have not a little contributed to its advance. Possessing, however, so many advantages, all other merchants that can, must in time follow them; unless the laws interpose in favour of the middle ranks, and of the country in general; more especially as those merchants who have not factories, are ever more and more harassed and disgusted, with the bad, falsified, and dear goods, now, by reason of the dearness of wool, exhibited by the old usual makers, (who are nevertheless nearly starving), and know not where to sell them with any prospect of profit or comfort.

We shall then have our valuable middle classes swept away, by the monopolizing power of the rich merchants; and this great trade conducted on a system, which, to it, is not at all needful, and which it would be desirable in a moral view to avoid in all trades, if possible, viz. *the large*

factory

factory system: Bringing together hundreds of men, women, and children, from their happy domestic labouring parties, and simple, uncontaminated manners, to nurseries always of vice and corruption, and often of disease, discontent and disloyalty. It would doubtless be well if the Legislature in its wisdom could devise some means of putting a stop to this needless, pernicious system in the woollen trade, pregnant with so many evils to the morals and well-being of this part of the country.

If encouraged to the utmost, no more cloth in the course of a year, could be made—it might be done with more dispatch, and the year's wool wrought up in nine or ten months suppose; but admitting a real demand equal to this dispatch, which is not the case, and that the foreigners would take it as quickly as manufactured; why for the sake of accommodating them a month or two earlier, but doing in fact at the long-run no more business, should we tread down great numbers of our own men, and set up more machinery for the purpose? It is still greater folly, when this is done without a real demand keeping pace with the mills, which has long been evidently the fact. This is an evil, however, which one would think would soon correct itself; but year after year we have seen that it does not, and that woollen goods are furnished much faster than there is any real solid demand for them, and therefore hawked about in almost every country; when, by a better arrangement of the trade, particularly by manufacturing with less dispatch, wool would be kept at a moderate price—the manufactured articles good and cheap—they would be every where greedily sought after—no manufacturing losses be incurred—and we might bid defiance to the utmost malice and power of our rivals, who will soon, no doubt, be in possession of all our mechanic arts; but let us only get our wool as before; it will then be so much lower and superior to any foreign wool, that with that, and our capitals, we might be tranquil respecting the issue, of their use of machinery.

It seems perfectly evident from what has now been said, that we have already too much woollen machinery; that the excess does not correct itself very readily, because the very thing that does the mischief, is unfortunately regarded as the cure; that we manufacture by far too rapidly; that great losses are somewhere constantly accruing in consequence, when, by less dispatch, every one might be well paid for his labour. Our great object consequently should be directed not to quicken by more machinery, and increase the evil, but by every practicable means to retard, and thereby decrease the intolerable dearness of wool.

Instead

Instead therefore of repealing all old statutes, with a view still to facilitate the quick making up of wool, more restrictions, rather, seem necessary, or the old ones repealed and new ones enacted, better adapted to the spirit of the times, and particularly, that the apprentice qualification should be somehow continued, as the best and surest way of keeping the number of clothiers and dressers within some tolerable bounds, and proportioned to the quantity of wool produced. *And above all, that full consideration be given to the large factory system,* and if found unnecessary and fraught with the mischiefs here represented, as no doubt it will, on due investigation, let the merchants be confined to their own profession. By such a measure, essential service would be rendered to the middling men of every description, and the morals of our labouring classes preserved, without any loss or detriment to the trade in general; since the wool, being the same, after, as before their erection, whatever is done in those factories is necessarily drawn from the old middling makers above described, who themselves make use of more machinery already than is prudent: But they will, notwithstanding, persevere, thinking they must have still more and more, in order to make head against the ever-rising price of wool; not suspecting *that,* to be the natural effect of their very efforts to keep down the price of the manufactured

factured article; and not reflecting that the sooner they get a piece made up, the sooner they want wool to begin another. Wool thus called for cannot but rise.

Another cause which prevents a ready correction of an excess of machinery is, that every new adventurer and mill-builder, flatters himself with more skill or luck than his unfortunate neighbour, and the unprofitable distressing round is again gone over.

See below the prodigious number of pieces manufactured for the last ten years in this riding, according to the annual reports: A considerable portion of which, as is indeed no wonder, have been, it is well known, sold without any profit to the manufacturer. It is also almost as well known, that after all this amazing quantity done, the clothiers as a body are impoverished—in a worse condition, and less at ease in their circumstances, than when considerably fewer pieces were made, and less machinery used. Their situation therefore is highly worthy the attention of Government. They labour hard, toil incessantly, suffer patiently, and know not how to help themselves. The Legislature alone seems competent to grant them relief and assistance, and to protect them, as well as the cloth-workers and shearmen, from falling a sacrifice to the spirit of monopoly—to private cupidity in the guise of public good.

A STATE-

A STATEMENT

OF THE

QUANTITY OF WOOLLEN CLOTH MILLED IN THE WEST RIDING

OF

THE COUNTY OF YORK,

WITHIN

The last Ten Years, ending April the Sixteenth, 1803.

YEARS.	PIECES.		YARDS.
1793,	Broad, - -	214,851 - -	6,760,728
——,	Narrow, -	190,468 - -	7,531,698
1794,	Broad, - -	190,332 - -	6,054,946
——,	Narrow, -	150,666 - -	6,783,722
1795,	Broad, - -	190,988 - -	6,067,208
——,	Narrow, -	130,403 - -	4,634,258
1796,	Broad, - -	250,993 - -	7,759,907
——,	Narrow, -	155,087 - -	5,172,511
1797,	Broad, - -	246,770 - -	7,830,536
——,	Narrow, -	151,594 - -	5,245,704
1798,	Broad, - -	229,292 - -	7,235,038
——,	Narrow, -	156,709 - -	5,503,648
1799,	Broad, - -	224,159 - -	7,134,114
——,	Narrow, -	148,566 - -	5,180,313
1800,	Broad, - -	272,755 - -	8,806,688
——,	Narrow, -	180,168 - -	6,377,277
1801,	Broad, - -	285,851 - -	9,263,966
——,	Narrow, -	169,262 - -	6,014,420
1802,	Broad, - -	264,082 - -	8,699,242
——,	Narrow, -	137,231 - -	4,833,534
1803,	Broad, - -	265,660 - -	8,686,046
——,	Narrow, -	137,016 - -	5,023,754

It

It also appears, from the above list, that notwithstanding we have every year more and more machinery, and the last a year of peace, fewer pieces are latterly made than before; the reason of which is, doubtless, that the old stocks of wool are gone, and we are now come to the real produce of the year.

If we wish to produce a good, cheap, and every where marketable woollen article, we must depend much more on cheap raw materials, than on machinery.

FINIS.

EDWARD BAINES, PRINTER, LEEDS.

THE SPEECH OF RANDLE JACKSON, ESQ.

ADDRESSED TO

THE HONORABLE THE COMMITTEE

OF THE

HOUSE OF COMMONS,

Appointed to consider of the

STATE

OF THE

WOOLLEN MANUFACTURE

OF

ENGLAND,

ON BEHALF OF

THE CLOTH-WORKERS AND SHEERMEN

OF THE COUNTIES OF

Yorkshire, Lancashire, Wiltshire, Somersetshire, and Gloucestershire.

PUBLISHED BY THEM FROM THE SHORT-HAND COPY OF

MR. GURNEY.

London:

Printed by C. Stower, Paternoster Row,

AND SOLD BY J. STOCKDALE, PICCADILLY.

1806.

THE

SPEECH,

&c.

MR. CHAIRMAN,

SIR,

THE period has at length arrived when it is become my duty to review the case of those petitioner's whom I have the honor to represent, and to offer to the committee such observations upon the evidence as may best elucidate their claim to its just consideration.

In discharging a duty of such extent and importance, I feel extremely anxious to abridge my address as much as I can, consistently with the interest of my clients, and to avoid as far as

I have strength of mind or memory, troubling the committee with a single superfluous observation.

I know not how better to treat this case than by supposing that the parties before you on the one hand are the masters, and on the other the men; and that those to whom I have the honor of addressing myself, are seated as judges between them. It is indeed necessary that I should have the liberty so to consider the parties.

It will occur to you that such would be the course of proceeding in ordinary cases before Parliament, as where a bill is to be supported or opposed. The present is a committee of inquiry appointed to collect information; and the nature and necessity of the case has imposed upon them the task of interposing such interrogatories as under other circumstances would have been done by counsel, unless, therefore, I have the liberty of considering you as judges, I should be deprived of that freedom of speech so absolutely necessary for me to enjoy in conducting the case before you.

I trust, Sir, you will do me the justice to believe, that I am not less anxious to discharge my duty upon this occasion in a manner acceptable to yourselves, and that nothing would give me more concern than to be thought for a moment to trespass those bounds of deep respect which I entertain for the committee. Should I in the course and earnestness of observation be so unfortunate as to raise such a suspicion, I entreat you before-hand to be assured that it will be unfounded, and totally repugnant to the feelings of my mind.

Sir, when I stated the committee to be seated as judges in this cause, I did it under the impression and conviction that you will feel yourselves called upon in that character to appreciate the evidence before you by those rules which courts impose upon themselves; for, however you may have varied from them in the form of your interrogatories, or in your mode of obtaining information; yet as judges between the parties, you will not feel yourselves at liberty to depart from that principle of immutable justice upon which the law of evidence, its construction, and application, is unvariably founded.

Meaning then to observe upon the evidence before you, as if it were put in before a court, permit me to notice some of the rules to which I have alluded.

One of the rules for the interpretation of evidence is, that when witnesses are not disparaged by their character or by their conduct, they are entitled to equal credit without regard to their stations in life. Another is, that when witnesses give contradictory testimony without standing impeached by either their character, conduct, or situation, numbers are allowed to prevail.

These are the principal rules as to personal credit, and I have no hesitation in saying, the committee would depart from its duty, were it to concede or to withhold belief by any other rules than those by which the same man would or would not receive credit in a court of justice.

Sir, you will also be enabled to get at truth by another rule, to which courts are never inattentive; that is, where it happens that either party, in the endeavour to establish particular points, betray themselves into certain admissions, in such cases, judges and juries ever fasten upon unintentional admissions with more earnestness than upon any main proposition that falls from the mouth of the witness, and give them greater weight from its being evident that they cannot be premeditated.

With these preliminary observations as to the nature and the mode of construing evidence, I shall take the liberty, before I state the different heads of the case, to dispose of one subject which has arisen out of it; it is one that has become of no inconsiderable importance, from the way in which it has been treated, though I trust that ultimately it will not be mixed with the cause of the petitioners; but if it be reported to the House of Commons, will be offered as a distinct consideration—I mean the Institutions. I beg leave to state, that whether such Institutions consist of masters or of men, they are not in themselves illegal. It is before you in the evidence of 1803, and in the evidence of the present session, that both masters and men have had their meetings, have assumed the same delegated form, if that signified any thing, and have associated together for the purpose of promoting their interests in their respective departments of the trade. I again beg permission to say, that while *Magna Charta*, and the Bill of Rights, shall remain, Institutions of that sort will remain unimpeachable. Englishmen have an undoubted right to confederate to maintain themselves against proceedings unjustly commenced, to prosecute others who may offend against their legal privileges, to support themselves before the legislature of the country, or to oppose those who seek enactments which they may think threaten their prosperity. I repeat, that Institutions for such purposes are most legal. It would be ridiculous and absurd to talk to the subject about his right to petition—to tell him of the consideration of parliament, and that the ears and hearts of its members are always open to the petitions of the people, if they were not allowed to confederate, and to apply their money and their talents in the exercise of that right, for the purposes of bringing their case fairly before parliament, and maintaining it when so brought.

We find throughout the evidence in 1803, that almost every master that was called, for I put the question to every one of them, was a member of the committee of masters; one of them was the chairman of their Institution: they all admitted that they subscribed for the purpose of carrying on parliamentary proceedings on one side, as did the men on the other; and both of these institutions I maintain were strictly legal.

I have, however, explained to my clients in private, and I wish them now to hear from me in public, that though they have a right to associate for such purposes as I have mentioned, they have no right to apply any part of their funds to the support of men who may have agreed to strike work. Neither have the masters a right to do what they have done; namely, confederate together with a view of lowering the wages of the men, or of agreeing with each other not to employ any man concerned in this defence before parliament. Both masters and men were wrong when they carried their Institutions to this length. It might, indeed, become a question if the machine which they are said to have combined not to work after be *an unlawful* machine. But I will not for obvious reasons touch upon such distinctions; I would rather they did understand, and I believe they do understand, that combinations not to work, or to sustain others in such a resolution, are generally illegal. I think I can hold myself personally responsible for every guinea and half-crown that shall in future be given to men to encourage them to forbear from work; I think I can pledge myself to the committee, that it is the last offence of this description that they will hear of from that order of persons whom I at present represent. Their error in this respect has been casual and inadvertent; they have misconceived the law, they know better now, and I think I can venture to say that the mistake will not be repeated. But I need not say to gentlemen like you, experienced in the world, and habituated to the thoughts, feelings, and I had almost said the prejudices of parliament, that if the subject of the institutions, described as I have heard them in terms of terror and offence, should be mixed with the rest of the case; and if the house should not be enabled to get at the state of the woollen manufacture, and the important facts which concern the gig mill, the apprentice's law, the searchers and sealers, &c. without perpetually stumbling upon it, it must have a tendency to make an impression of the most unfavourable and unwarrantable nature in the minds of those whose duty it is to be strictly impartial. I therefore make it a claim of justice, as well as of policy, and of wisdom, that in whatever report you make to the house, that subject shall stand by itself, and not be mixed with other matter, so as to affect the evidence necessary to inform the houses of parliament as to the true nature and state of the manufacture of woollen cloth.

I should disguise my feelings if I did not say I entertain no great apprehensions of any law taking place upon the subject alluded to severer than that which at present exists; if there should be a revision of that law, I think it would produce one infinitely more mild than the Act of the 39. and 40. Geo. 3. I am not here denying the necessity of that act, but it is impossible for any man who knows and loves the constitution of his country, not deeply to deplore its necessity, if necessary it was. I have no fear, I say, of a law stronger than that being brought into parliament, because no stronger one can be passed unless it deals in blood; it will then produce the consequence which all severe laws do, namely, non-enforcement. I trust, therefore, that the subject of Institutions will be kept entirely distinct from those branches of the case which I enumerated in my opening address, and to which I shall now proceed to speak.

The points in question are, the gig-mill, the sheering-frame, the searchers and sealers, and the apprentice' law.

With respect to the gig-mill, much has been said as to machinery in general, and much evidence has been gone into, indeed, to state the extreme impropriety of checking machinery. I shall, therefore, beg leave to say a very few words respecting machinery in general, as applicable to those particular machines, to the toleration of which we object, and which are already prohibited by law. I cannot be insensible that there is no one subject which our countrymen have taken up with more unqualified enthusiasm than the application of machinery to manufacture. If there be one upon which more than another their ingenuous tempers have run wild, and respecting which they have been more in error, it is that of the indiscriminate application of machinery. They will not deliberate upon those due distinctions which the best writers have made between the use and the abuse of machinery. I shall prefer, upon this occasion, the giving you the sentiments of eminent men to the obtrusion of my own; first laying down two or three general propositions.

I am perfectly aware of the stupendous advantages which our country has derived from machinery, but it does not follow, therefore, that its unlimited use would be wise; on the contrary, we know that machinery by its supercession of human labour has a most depopulating effect—to use the language of political economy, it deprives the land of so many customers: suppose, for instance, that by means of a machine you should be able to transact as much business with one thousand men, as if done by hand would require ten thousand? The land is in that case deprived of nine thousand customers, there being of course a proportionate decrease in population, and therefore in cases where the advantages of machinery do not overbalance so great a political evil, it ought not to be encouraged.

Now in what way is it stated by those who have thought most upon the subject? The wisest men acknowledge the difficulty of drawing the line; they all admit the first evil of machinery, namely, its proportionate depopulation, and the consequent loss to the land of so many customers or consumers. Unable, therefore, to fix precise limits, they lay down certain general rules; as for instance, that unless machinery be necessary to enable you to meet competitors at a foreign market, you should not hastily encourage it, because if you have the foreign trade in your hand, as we have in our case; if you are unrivalled, as it is admitted we are in the woollen trade; if no fault be found with your prices, which is the case with us, and that even the goods manufactured by machinery are sold, as has been proved to you respecting the gig-mill, as dear as the goods wrought by hand, and are not pretended to be better in quality; then you have none of those high advantages which should alone induce you to risk the evil of depopulation.

I will read a short extract from a work of great reputation, because it generally comprises the thoughts of the ablest men. In Postlethwayte's edition of Savary's Dictionary, it is said, under the titles of Wool, Cloth, and Machinery:

"The increase of our woollen manufactures will keep up the price of the product of the land, as demands for goods not only employ our poor, but cause the advance of the price.

"The products of land in England is a great part of every manufacture; our rents are but the value paid for the products of the lands, therefore all the additional labour we encourage, which pays the product of the lands, is so much added to the rents of the kingdom.

"An increase of wealth may attend a decrease of numbers and real strength, for which reason the balance of money is less to be regarded. The exports of our nation may at present exceed a million in the lower branches of baize, serges, druggets, and flannels, and this may employ a *million* of hands. In a course of years, we will suppose this trade to be changed for that of wrought works, tapestry, painting, and statuary, in which our exports might amount to *two millions*, but for which a *thousand hands* full employed would be more than sufficient.

" Here we see it possible that our wealth may be annually increased *a million*, whilst our *real strength* is decreased in the proportion of *a thousand to one*. There is nothing incompatible in the arts of ingenuity, and those of labour: and all the plainer trades may be retained, notwithstanding the higher arts are introduced, in which case we should be both a richer and greater people.

" In such a variety of reasons it is not easy to determine to what degree the use of machines in general should be admitted. States without commerce regard chiefly the increase of numbers, and their home markets; and commercial states are too apt to consider wealth alone, and foreign markets;—machines, without prejudice to either, may be allowed in the following cases.

"When they do such business as cannot be performed by hand at all; of this kind are pumps, fire-engines, water-engines, looms, wine and oil-presses, hand-mills for grain, and, perhaps, horse-mills.

"Where the commodities wrought by them are such as would not be used at all, except they were done by the machine, either being not cheap enough, or not good enough, for consumption, when prepared by hand. Under this head are the mills for making paper, those for forging and slitting iron and copper, fulling of cloth and leather, and making gunpowder.

"A people without commerce may safely refuse to admit stocking-looms, sawing-mills, throwing-engines, weaving or spinning-engines, mills for striking files, cutting watch wheels, making nails, and all the variety of inventions produced by a rivalship amongst nations contending for commerce, and private men for orders.

"Commercial states must have their eyes

upon their neighbours, and if they design to engross foreign markets, must provide for the cheapness of labour at home. Goods must be made cheap to render them of general use abroad and at home. Engines for shortening business ought to be rejected, or not admitted in commercial states, when the commodity is not all sold abroad; when it affects not the price of labour, and *when the machines would lessen our home markets more than increase our foreign ones.*"

It will be admitted, certainly, upon the authority of this writer, at least, that machinery should not be used with too much freedom, unless those necessities exist which he describes.

Montesquieu, vol. ii. cap. 16, page 290, says, "those machines which are designed to abridge labour are not always useful: if a piece of workmanship is of a moderate price, such as is equally agreeable to the maker and buyer, those machines, which would render the manufactory more simple, or, in other words, diminish the number of workmen, would be pernicious.

I will trouble the committee with another extract.

M. de Boulainvilliers, in his view of the commercial interests of France, entitled, "Les Interests de la France mal Entendu,' tom. iii. p. 272, enters more particularly into the subject.

"A malady," says he, "is expanded over our arts. I would speak of machines which tend to simplify and diminish the labour in our manufactures. The cabinets of our ministers are every day filled with projects for machines proper for lessening the number of hands employed in our fabric. From hence it results, that a multitude of artisans in France are become useless, and must offer their industry to others. I do not speak only of springs and movements, all the manufactures of the kingdom will soon form themselves; they will be in no want of artisans. The *first object* of manufacture is employment, or, what is the same thing, of giving subsistence to a great number of men—wandering from this principle is contradicting the institution itself of the arts, it is diminishing the number of men, for every time that *we place bounds to subsistence, we do the same to population;* so established is our prejudice, and so generally is it received, that we are come to confound the produce of industry with the means of subsistence, which ought to create industry, two things very different in their principle and object. Let us suppose a manufactory to employ 10,000 citizens, and on the other hand, 500 machines, which produce the same amount of labour, in the same art. It is certain, that the first gives a living to a much greater number of subjects, and therefore completes a much more important object. The maintenance of ten thousand artisans, who draw their subsistence from a branch of industry, forms, in the state, a multitude of other smaller manufactures. But to set the inconvenience of this prejudice in the stronger light, we must begin by curing another:—I speak to those who establish it as a principle, that when a machine diminishes the artisans in one art, that they directly apply themselves to another. Since the taste of mankind is so very refined, and that one great luxury always succeeds another, the arts become so complicated that a man has not time, in the course of his life, to master above one. If he loses that by any accident, he can never have any hope of replacing it, if it were not so, we should not find, in every revolution which happens in an art, so many idle men, and so great a number of useless subjects. It is objected (and here lies the strength of their system), that machines in diminishing the price of labour bring in great riches to the state, but they do not bring into their account the precarious manner in which these riches are distributed ; they concern only a small number of particular persons, *the proprietors of the machines.* They cannot demonstrate geometrically, that a sum of ten millions which circulates among fifty thousand manufacturers, is better for the state than one of an hundred which circulates amongst a thousand."

We have here the sentiments of an able and intelligent Frenchman, holding a high station in the political service of his country; his design was to check the inordinate propensity which then existed towards machinery; he also lays down certain rules as to the wisdom of its application; and those rules may be summed up in the two or three propositions which I have before had the honour of stating, namely, That when machinery is not necessary for the purpose of lowering the price of the commodity—when it is not necessary to be used because you cannot make the article by hand, which is the case with some articles—when it is not called for in order to produce the article of a more excellent quality than you could render it by human labour—and, when it is not necessary to be used in order to succeed in the foreign market; in these cases you should, according to the writings and opinions of every great man who has treated upon political economy, interpose your authority, and say, so many customers shall not be taken from the land, unless to procure such advantages for the public as will counterbalance their loss.

In the case before you, it appears, from the testimony of the factory-masters themselves, as well as from that of every other adverse witness, who have spoken to that part of the subject; that, independently of the machinery in question, we have been able to complete at foreign markets to such a degree, that we have the trade entirely in our own hands; if so, you are not called upon to inflict that evil upon the landed interest, which in so great a degree, attends the depopulating effects of the gig-mill and shearing-frames.

I have stated this generally, but if I shew

that the particular machine in question, viz. the gig-mill, while destructive to the best interest of my clients, is absolutely injurious in its nature to the fabric, and that it is a machine not newly introduced, but the same which was attempted as far back as the days of Edward the Sixth, and then prohibited for those faults which characterize it now, and are inseparable from it; then my case stands upon the highest ground that it is possible to place it, for not only will it have been shown to be a machine of the most depopulating tendency, and, therefore subject to the general and positive evil alledged, but proved to be as injurious in its application to the fabric now as it was when prohibited by the statute of Edward the Sixth.

It cannot be said that it is not the same machine. Mr. Law Atkinson, like a man of sense and candour, admitted in 1803, and has admitted again this sessions, that it is the very same. An action was tried before Lord Loughborough some years ago, as to its identity; Mr. Atkinson, whose uncle, I believe, was defendant in that cause, says the cause was lost from a defect of proof, but that he never entertained a doubt of its being the same machine as that which is prohibited by the statute of Edward the Sixth. He adds, indeed, that it is prohibited for unwise reasons—I say, for wise ones; and therein we differ—I having made out, most completely in evidence, that the gig-mill is as injurious in its operation as it is baneful in its effects.

In an address of this nature I can only refer to evidence generally; and I contend that, you had in 1803, and have now, a stream of such testimony as according to the rules of construing evidence must weigh, such, as according to every one of those legal maxims by which courts and juries govern themselves, in its application cannot but prevail. I mean, that it has been proved by the testimony of men who have personally worked at the machine---who have had experience of it in every stage—who have repeatedly examined the cloths both before and after its operation, that it has a tendency to strain the cloth to an infinitely greater degree than when it is dressed by hand. When cloth is dressed by hand it seems it may be stretched or not, according to the choice of the manufacturer, but when dressed by the gig mill it must inevitably be stretched beyond a fair and fit degree, except as to some particular cloths, such as I shall presently notice, which require very little dressing, and are drawn but a very few times through the mill. It has also been incontestibly proved, that the overstraining produces those injurious and deceptive effects upon the cloth which the witnesses described in 1803, and which were described, deprecated, and forbidden in the days of Edward the Sixth.

All cloth, in proportion as it is attenuated, they admit on both sides, will feel soft; it yields a nice and delicate sensation to the touch, it appears much lighter and will have a greater degree of gloss than when it is done by hand, because it is then less stretched; and I have no doubt but this is the deceitful making assigned as a reason for the statute which prohibits the gig-mill. It will be recollected that the witnesses on the adverse side, when denying these effects charged upon the gig-mill, referred to a kind of cloth which does not come within the argument, it not being capable of distention—viz. friezes, blanketings, and coarse cloths, which they allow, are put *dry* upon the machine; and it is a proposition of natural philosophy, that what is put on in that state will not stretch, it is only when damped or wetted that the cloth will extend; these friezes, blanketings, and other coarse cloths, are strained by a previous process, and, as it appears, to an amazing length; but it is in evidence they do not receive their extension upon the gigtmill, they being put on in a dry state.

The balance of evidence has incontrovertibly proved that the superfine cloths, and the mixed medley cloths of various descriptions, acquire a considerably greater degree of extension than is fit and suited to the fabric. I will refer you to what has been said by the masters themselves, begging leave, first, to recal your attention to one of the propositions with respect to evidence with which I set out, namely, that when *admissions* escape from the adverse party, they are always entitled to the greatest possible degree of attention. This observation will particularly apply to the evidence given by the masters in 1803, which is before you as a part of the case, as well as to that given by them this session.

Mr. E. S. of Uley, in Gloucestershire, admits, upon his cross examination, "that cloth *may be overstrained*, so as to afford *an extraordinary profit* at foreign markets, *without breaking the thread;* that great art is required in using the gig-mill, and that the cloths done by the hired or public mills have been done badly." He admits, you see, that a certain degree of extra straining may take place without breaking the thread, in which state the cloth is pressed and goes to the customer without the damage being discoverable, or if discoverable, only before a strong light by those whose trade it is to examine it; and even, if when the ground is broken, we have abundant proof that it is the custom in those cases, so to fine-draw, flock, and press, as almost to defy detection. But I quote the evidence of this master principally to shew that a great degree of illegal stretching may take place *without breaking the thread,* and without any injury to his interest to be compared with the advantage which he derives from the amount of the cloth so extended.

Another gentleman, a Mr. W. of Dursley, in Gloucestershire, states, upon his cross examination, that his observations are confined to the strong cloths of Gloucestershire. I notice this observation, because if the committee should consider whether they cannot introduce some such qualifications in this business as may heal the unfortunate divisions which prevail; they will inquire to what cloth the gig-mill can be safely applied, and to what it cannot: some line of demarcation of that kind would perhaps be the wisest course that could be taken for the fabric, and would materially tend to restore peace among the manufacturers. This gentleman, (Mr. W.) allows it to be his opinion, that the medley cloths dyed in the wool will not bear *the pressure of the gig-mill*, like the white cloths of his county; and he goes on to say, "that they leave their lists longer on the cloths meant to be dressed by the gig-mill than on those dressed by hand, *to provide for the necessary degree of straining in the gig-mill;*" he adds, that "some clothiers may stretch them more than his concern does; that the great advantage of gig-dressed cloth is in its *appearance*, and that they sell such cloths for a *higher price* than those done by hand."

Another gentleman, Mr. E. a clothier of Chippenham, admits, on cross examination, that "the cloths done by the public mills are improperly done; when bracks are made they are fine-drawn and *flocked*, and that if the cloth be more strained *than is usually done by hand*, it feels *softer*." This latter quality you have been informed will be in proportion to its straining.

Mr. William S. of Frome, in Somersetshire, allows, that "they mill their cloths one yard short in an end, to *bear the force of the gig-mill.*" This, you will observe, is at least, an admission of its greater force, and so far confirms the workmen, who say, that the ordinary application of the gig-mill strains, at least, two yards upon twenty beyond the legal length.

Now as far as admissions from adverse parties have any weight, and it appears to me that they are always entitled to great weight indeed, they make out my case for me, namely, That the mill, in its nature, produces an infinitely greater degree of pressure and straining, and that in its application it has produced much extra illegal and improper straining upon most of the cloths dressed by them, which though not so injured as immediately to hurt the character of the manufacture in the foreign market, yet must have been so made from the nature of this machine, as, by and by, will have a very calamitous effect upon the interest of those who are obliged to follow the trade of a cloth-worker for the maintenance of themselves and families.

I said that I would give you some idea

of the depopulating tendency of this machine. It is admitted that three persons, with the assistance of the gig-mill, would dress about as much cloth as twenty-four persons could dress by hand, and that calculation was made to shew how immense a saving it would be. We must undoubtedly allow that, if the gig-mill will save so much hand labour, it is an irresistible argument in its favour; if not met by the propositions which I before stated; but you will not deprive the land of twenty-one customers out of twenty-four; that is, you will not throw them out of bread unless it be for an adequate political advantage. In this case, I contend that it would not be wise or just so to do, because the gig-mill does not do the work better, because it is not necessary for the maintenance of our preference at foreign markets, because it has not rendered the cloth cheaper, and because our customers are at present satisfied;—in a word, because it has all the disadvantages which political economists decry, without being attended with one advantage sufficient to justify a system of depopulation.

Had we no other proof of its depopulating tendency than merely the evidence with respect to the earnings of the unfortunate persons in the West of England, where this machine has been principally introduced, it would be sufficient. I have in my hand a calculation, stating their average earnings for the last four years and a half—it is but about fourteen shillings a week; while, in the north, where the mills do not prevail, the workmen get a guinea and a half. Can any thing more completely prove that there are persons ready, and in number sufficient to do by hand all the work that can be required of them, then the fact of good workmen being thus attainable for so scanty a pittance? this fact was stated over and over again, by some of the most respectable master clothiers of the West, in 1803; they said there was a plenty of men for every part of the woollen manufacture; indeed, the adverse masters proved a redundancy of hands, by gravely bringing a surveyor of the high roads before the committee to shew that he had offered work to the immense number of persons thrown out of employ by this fatal machine. He boasted that he had offered them work; that is, he had offered work to persons who had served regular apprenticeships to the woollen manufacture, in the hope of enjoying the privileges legally resulting from a faithful apprenticeship! And what work did he offer them?—why, that they should break the stones upon the roads, and cleanse them from filth.—That many canals being begun, they should fill barrows with the soil, and wheel it away! He added, that though he himself had offered them this employ, yet were these clothiers fastidious enough to say, they liked cloth-working better than this new and honourable employ of breaking the stones on the road!—When the gig-masters found it necessary to protect themselves from the charge of turning the men out to starvation, they thought it necessary to shew by this witness that they had offered them this sort of work!—that they had offered it to men skilled in working the great staple of the land!—men who had acquired legal privileges by long and faithful servitude!—it is to such men, they say, that they have no right to complain of the calamitous consequences of the gig-mill, since they may find employ in breaking stones upon the road!—in cleansing them from filth!—in filling barrows with the soil, and wheeling it away!

These are the principal observations which I mean to make upon the gig-mill; I will, therefore, next in order, notice the sheering-frames. I shall have something further to say, of course, with regard to the illegal straining of cloth, when

I come to speak of further and recent admissions which appear upon that head. With regard to the sheering-frame, it is nearly, though not precisely, in the same predicament as the gig-mill. The gig-mill stands strictly prohibited by the law in the reign of Edward the Sixth, and those persons who seek to put down the gig-mill, are but of opinion with their ancestors: they only call upon you to enforce the law, which is entitled, "An Act for putting down of Gig-Mills:" and for the same wholesome and wise reasons, namely, to prevent the use of a machine which was declared to be injurious to the fabric then, and is proved to be so now. The sheering-frames, however, are not prohibited in the same distinct form of words, but they are prohibited in spirit and in fact; it is enacted, that no knife, blade, pummice-stone, or thing that shall have a tendency to sheer cloth uneven, shall be used, but that sheers only shall be used for that purpose. With respect to the sheering-frame, the machine itself was not known in those days, but its effect was known, deprecated and forbidden. The sheering-frame is unfortunately within the principle of the gig-mill, it having a tendency of the most cruel and depopulating nature, and is not called for by any one political reason that can justify its establishment.

You have heard, Sir, from the masters, one and all; and from many of the questions put by members of this honourable committee, it was plain they inclined to believe that although the gig-mill drives twenty-one out of twenty-four out of employ, yet upon the whole there has not been less employment for the men, it having been asserted that this machine manufactures so great an additional number of cloths, that the twenty-one thrown out of employ by the gig-mill out of every twenty-four, find refuge in the arms of the sheering trade; that is, that as it manufactures so much more cloth, so many more men are necessarily employed in the sheering of it; therefore, said one of the clothiers, though our sheermen objected to the gig-mill at first, they soon became reconciled to it, because they found that it more than compensated them by an increased quantity of labour in another way.

Another gentleman says he called his men together, and stated that fact to them, and they became at length convinced; and Mr. Law Atkinson says that his men have no objection to working after the gig-mill, because as the gig-mill raises so many more cloths, it makes so much more sheering work for the very persons whose labour in another line it supersedes; and they get so much money on that account, that they are become friends to the gig-mill, that must mean, upon the supposition that the whole of the sheering would be done by hand.

Some consolation of that sort was administered by the evidence of the masters in 1803, and we had even taken a little of it and began to be comforted, when at length came these sheering-frames to swallow up nearly all the manual labour that the gig-mill had spared, for it seems that the twenty-one men out of twenty-four which have been driven out of employ by the gig-mill, and for whom it was pretended the extra sheering would find work, are now to be altogether undone and destroyed by the introduction of frames to perform that task. Is it wonderful that men should sink under these accumulated calamities? Can you be surprised that the thousands and tens of thousands of the petitioners, and their innocent families, should weep and despair under distresses of this kind? I must not be told that in Yorkshire they get from a guinea to a guinea and a half a week, and therefore they should be satisfied. So they do:—but how do they get it? It has been proved that it is by working from four in the morning till twelve at night; be it so, save them but from the peril of these two machines, and they will not repine at these laborious exertions. In the other parts of England, they are not getting more than fourteen shillings a week; the sheering business which was to get the men into conceit with the gig-mill being itself about to be done away by the sheering-frame, which comes to complete the work of desolation, and to deprive of their remaining labour those hands which were left unemployed from the gig-mill.

I think I can show reasons of the most valid kind against the use of this latter machine. I will not rely for those reasons on the body of evidence which my clients have adduced; but I will, in spite of themselves, draw proofs from the mouths of our adversaries. I will invoke the opposing witnesses, and shew from their own inadvertent admissions, the utter impolicy of using this machine, supposing even that it had none of those cruel and depopulating tendencies with which I charge it. It was stated by the evidence, in chief, who assigned grounds for their judgment, which there is no controverting, because they are natural, and speak for themselves, that the sheering-frame has this mischievous effect, it will not *sheer even,* and if the cloth is not put on perfectly even, it will knib, that is, catch it with the blade. Now, if it knibs when they sheer by hand, the sensation of the sheerer immediately detects it, and the injury is stopped. The fine-drawers have said that the holes made in the cloth by the sheers are of the size of a pea, or a gun-shot; while, on the other hand, those made by the sheering-frame are generally three or four inches long: the reason is, that when a hole is made, it cannot be detected till the blade, which is some inches broad, and fixed in the frame, has passed over a space equal to its breadth; they then find out that a hole is made, and they stop the machine; but those men who have worked at it themselves in person, and those who have looked over the cloth immediately after, tell you one and all, that such is the nature of the machine. This evil cannot be prevented, however ingeniously the fine-drawer may repair it.

To the evidence of these persons is opposed that of a Mr. B. who describes himself as a very profound man in his way, an expert mechanist, and one who says he has passed all his life among machines of various natures. I am afraid if it were a question of moral character, we should have little to say to this person upon his own showing, because it seems that after living fourteen years with the traitor Douglas, the fellow that first attempted to get our woollen machinery over to America, and who is now endeavouring to introduce it into France, and is there acting a part that ought to bring him to the gallows, if he could be laid hold of, he favored this Mr. Douglas with his drawings; the better to enable him to transport our staple manufacture to America; I will, however, forget that, let him stand as fair before you as I trust all my witnesses do, and attend only to his evidence. He says the sheering-frame sheers more even than if done by hand. Why? Because, adds he, one motion governs the whole, and after the machine is set, it cannot vary, according to the eternal laws of nature, from that mechanical motion with which it commences its operation. This, he says, is as true as any proposition supported by mathematical demonstration; he says it must have that effect, because when once the machine is fixed, as to its due poise, it goes on and sheers *even*: well, but suppose the cloth should be previously *raised uneven*, how does it do then? for if the machine cannot vary, if it cannot go to the right or to the left when the cloth comes uneven from the gig-mill to this machine, as is proved to be frequently the case, the effect must necessarily be to cut it most injuriously from that very undeviating regularity of its motion, which Mr. B. speaks of as its great perfection.

Now, the sheermen have the cloth placed before them, and if one part of it be balky in the ground, and in another part the nap is scarcely raised, or if the cloth be in any way uneven, they govern their sheers accordingly, they move them lightly over the balky place, and press them more forcibly where the nap is left too long. The sheering frame cannot, in the nature of things, be so managed; because, as Mr. B. tells you, one motion governs it from the beginning to the end.

It has been admitted by the masters, that the greater part of the cloths are *raised* at public mills, and that public mills raise the cloths uneven, and it is most clear that every cloth which is *raised uneven*, whether from ignorance or accident, must of necessity be injuriously cut by the sheering-frame when it comes to the uneven part of it. The regulating power which has been boasted of by some, even Mr. B. was not prepared to affirm, when I asked him if he could tell if a hole were made in the cloth by the sheering-frame till it had passed over, and reminding him of what had been said by the sheermen: viz. that they could *feel* instantly as the sheers in their hands knibbed. We make it out, says Mr. B. partly by *the sound*—the sound of a hole! What, says an honourable member, does the hole sound before it is made? This question put Mr. B. a little to his trumps, he gave up his exquisite sensation of sound, and then said, that if a person were extremely watchful, and kept his eye upon that particular spot, and saw that a hole was made, he could stop the machine before it got to be three or four inches, or whatever might be the breadth of the blade. Now you will recollect their boast to be that one man looks after half a dozen of these machines, for in that consists the great saving of labour. They say it is not necessary to have a man after each machine, as you have after a pair of sheers, but that one man superintends half a dozen of them; then I should be glad to know how, even with Mr. B's. faculties, a hole is to be discovered instantaneously, as is the case when a man sheers by hand, who can tell by sight as well as by sensation, the moment the sheers knib or catch the cloth.

My humble argument then is this, if the sheering-frame is liable to the legal objection made in early days against *sheering uneven*; if it is not called for by any political necessity; but that on the contrary, while it swallows up the whole of the labour left from the desolation of the gig-mill, it operates injuriously to the fabric: you will surely think it right to prohibit, by name and description, that which already stands prohibited by the law of the land, as to its principle and effect. I find that this is a machine which the masters are less disposed to persist in the use of than any other. Many masters that have been before you, state that it is but very partially used. Two of the masters state that they do not use it, they do not want it; and that if they could but have the free use of the gig-mill. all their cloths should be sheered by hand. Indeed the evidence of the fine-drawers is at once a proof of the mischievous nature of the machine, and their own exquisite skill in repairing holes of such magnitude. Mr. Dyer, I remember said, expressly, he liked hand-sheering best. I notice those sentiments because I am convinced of the absolute necessity that exists for something like an amicable arrangement of this business, something like a softening of that high tone with which the workmen have been treated, and being warranted by the evidence in saying, that the prohibition of the sheering-frame, which would afford so much consolation and protection to the men, is a matter of comparative indifference to the masters, I earnestly hope and entreat, that some qualification of that kind may be suggested in your report to the house.

The general answer which I understand is given to the arguments I have offered respecting the gig-mill and the sheering-frame, is an answer which, I am sorry to say, is more and more coming into fashion. I know not how it is, but a sort of commercial jacobinism seems to have succeeded to the political jacobinism which lately afflicted and infested the the world; it is indeed urged with more silence, but it is moving with infinite subtlety, and I am afraid will be found almost as mischievous.

The modern cant is, that mens' own interest is a sufficient security for their observance of right, that the contrary will work its own punishment, and that every thing is sure to find its level.

Our ancestors, it seems, were all wrong—

even up to the 13th of Geo. III. we were in the woods. We have been from the days of Richard II. making laws respecting the woollen manufacture for men who want no laws; every thing, again they cry, "Will find its level." These new doctors answer all my objections by way of interrogatory, and exclaim, "Can you think a man would be so blind to his own interest as to send to market cloth that is not manufactured in the best manner? or that if stretching it beyond the length allowed by law, deteriorated the commodity, he would be so absurd as to do it?" Strings of questions of this nature were put to the masters, who have been gravely called to state to the committee their exalted ideas of the honor of the woollen manufacturers. This is like calling upon a smuggler for his opinion of the revenue laws. The masters whom the woollen laws were instituted to controul, are absolutely the only evidence called to prove the wisdom of dispensing with them, and to shew the folly and extravagance of our ancestors, who thought restrictions respecting this trade to be wise and necessary.

But I do not need to combat this proposition, for it defeats itself; you might as well talk of morals finding their level as manufactures.

Were men not as virtuous seven years ago as they are now?—Was not the powerful principle of self-interest as predominant then as it is now?—Was it not the interest of the master then to make the best possible article?—and yet the legislature, up to that period, or within a few years of it, thought it necessary to continue the care and watchfulness of preceding Parliaments, in order to insure a good fabric. The legislature, time after time, in different reigns, from the 13th of Richard the Second, up to the 5th of George the Third, has made law after law, for the better regulation of the woollen manufacture, till by a succession of statutes introduced, as new frauds or faults discovered themselves, this great staple has attained to the reputation which has produced our affluence. It is this high reputation which we are anxious to maintain. We, the great body of its artisans live but by the fame of the fabric, which cannot be destroyed without destroying us, when that dies we shall sink and be extinguished.

There is, I believe, a Bill now before the House, which I mention from the strictness of its analogy to the present subject; it proposes that the butchers, whom our foolish ancestors thought it right to prevent from flaying beasts wantonly and improperly, thereby injuring our leather-staple, should be left upon this subject to their interest, and their honour! It is not more than three years since the legislature thought there was something required besides this universal principle of self-interest, and they obliged the butchers to take off the hides with proper care, under certain penalties. The butchers, I understand, apply to be relieved from this odious obligation, and they argue that the state may rest safely assured that they will not hurt their own interest, by doing injury to the hides. But, Sir, in all cases we find the *immediate* interest operate more powerfully than the *remote*. This is a distinction that does not appear yet to have found its way into the new philosophy, though it is evidently that which has hitherto governed the legislature throughout the whole series of its enactments.

Are we any of us so insensible to experience as not to know that every deviation from moral rule is *impolitic*, to say nothing of its being bad and wicked in itself? and yet does every such departure proceed from our preferring a present indulgence to a remote interest!—These masters will get fortunes by the extra straining of their cloths, before this new philosophy, this emancipation from restraint, can produce its mischievous effects upon the fabric. What will be to them the loss of its reputation? they will by that time have accomplished their object of gaining speedy independence, and seated snug at their villas, perhaps, be laughing at the folly which listened to their arguments, which exchanged the progressive wisdom and experience of five hundred years for manufacturing sentimentality, and risked the high reputation of the great staple of the country, rather than the infallibility of a theory which preceding ages have treated with the scorn it deserves!

If the straining of cloth beyond a given degree had not been improper in itself, how came our ancestors so tenaciously to guard against it, and to have said, that though, in the course of the process of your manufacture you may perhaps inevitably, or occasionally, overstrain one yard upon twenty, yet beyond that one yard you shall not go. If all restraint is to be abolished, as contended for; if the masters are to be perfectly at liberty to stretch their cloths to what extent they can, let us apply that licence to the argument of self-interest, and inquire how far the latte· principle may be regarded as a sufficient security for that reputation upon which solely depends the wonderful extent of our foreign trade in this article. A very respectable master, who has a gig-mill, was called before you in 1803, the average of whose manufacture appeared to be about thirty pieces of cloth per week. The witnesses, you must recollect, have uniformly stated the extra straining of the cloth to average, at least, *two* yards in twenty above the legal measure. I took the trouble to calculate what it would produce to such a man if he got only *one* yard in twenty above the legal length, although the greatest part of the evidence states that *two* yards may be obtained without *breaking the thread;* that is, the manufacturer may get an immense deal of money between the usual degree of extension by the gig-mill, and that degree which it is said would be its own punishment, by inducing the rejection of the article. Allowing, however, but *one* yard upon twenty, above the legal measure, the maker would, by such extra straining, get upon the manufacture of thirty pieces per week, 1989 *l.* per annum;—the pieces run about thirty yards in length, a yard and a half upon thirty, is, of course, equal to a yard upon twenty;—a yard and half, at 17*s.* 6*d.* per yard, is 1*l.* 5*s.* 6*d.*

each piece;—thirty pieces, at 1*l.* 5*s.* 6*d.* is 38*l.* 5*s.* per week, or 1989*l.* per annum;—Can you then imagine any kind of doctrine more delusive than that men will neglect their immediate and personal interest, for a remote and general advantage,—that for the sake of the future reputation of the fabric at the foreign market, they would, when relieved from the present prohibition of the law, forbear a practice by which the immense immediate advantage of two thousand a year is to be obtained upon thirty pieces per week? Supposing the extra straining to be but half of what is stated by almost every witness—is this charge of illegal straining without further confirmation?—One of the Yorkshire master's, who is a respectable magistrate in that county, was called: an honourable member of the committee reminded him of the unfortunate circumstance which led to the loss of the Russian market; namely, that a Russian brigade, newly clothed with English cloth, happening to be exposed to a drenching shower, their coats shrunk into waistcoats. Mr. C. the witness, admitted the anecdote, and added, that they had the most exorbitant demands made upon them by the German factors and merchants, on account of extraordinary straining! Now, if the fact of overstraining was not become more notorious, and if the consciousness of it had not induced occasional abatement in the charges on that account, is it likely that such demands would be made? It shews the idea is spreading upon the Continent, that our cloths are more strained than they ought to be, and this may be attended, by and by, with the utter disrepute of our manufacture. If the bare impression of such a fact is capable of doing so much national mischief, what will be the effect upon the mind of the foreign purchaser, when he learns that Parliament has thought proper to take off every restriction, and say to the manufacturers, go on and strain as much as you please—settle it between yourselves and the Germans, Dutchmen, &c. the legislature will no longer interfere? If your own interest is not sufficient to restrain you from over stretching the cloth, in God's name strain it as much as you like, we will have nothing more to do with it, we know that every thing will find its level, and that you, at length, will be punished for your cupidity!—Are you prepared, Sir, for the consequence of such an idea getting abroad? Are you prepared for the conclusion which foreigners must, of necessity, come to, the moment that they know that you have withdrawn from the woollen manufacture all protective and prohibitory laws? Can you even hope that the loss of its confidence will not follow the annunciation of their repeal? As wise and intelligent men, you cannot but expect that those suspicions which the masters themselves have proved to be already alive upon the Continent, will amount to confirmed opinion, when they learn that whether a manufacturer shall overstrain his cloth one yard or six upon twenty is no longer a question of law, but a matter to be settled between the vender and the purchaser!

I hope I have now given something like an answer to this new-fangled proposition of self-interest being a sufficient check upon the manufacturer without the aid of the law. I have shown, that by departing from the existing law, he can get from 2000 *l.* to 4000 *l.* a year, without discovery. The masters themselves admit, as you will see in the evidence, that this moderate degree of illegal straining is scarcely within the possibility of being found out, after the cloth has been pressed and folded for sale; it does not even call for the fine-drawer, who is only wanted when the thread is broken or cut. This mischief, when the law is repealed, can have no corrective, except what it may receive five or six years hence, when it shall be found to have deeply affected, if not destroyed, the reputation of the fabric. The manufacturer may henceforth overstrain his cloth without its affecting his moral reputation; for the moment the law of repeal shall have passed, it will be a contest who shall strain most in order to get most. There will be no longer those little idle contests about yards and half-yards, but the race will be who shall extend the cloth to the greatest degree, without so palpably injuring it as to defy the skill of the hot-presser and the fine-drawer.

The effect of such conduct is already most serious, and had not a single master been called, and had not a single admission been wrung from them, I think that the evidence of Mr. Lambert must have been decisive of this case.

This person is a servant in one of the first houses in town. He is, by profession, an over-looker or examiner of cloths for the drapers; that is, when the draper looks out a certain quantity of cloths at the warehouse of the factor, they are sent to a person whose peculiar business it is to examine and report upon them. Mr. Lambert tells you he examined, at least, two hundred cloths a week for various drapers; that by far the greatest proportion of those which pass through his hands are dressed by the gig-mill; and that he pursues a mode which cannot fail of detecting the injury, if any be done, to the cloth. He begins with taking out the press, that is, getting off the gloss, he can then discern what injuries there are; and he tells you, that out of two hundred cloths he is obliged to throw aside a considerable number, sometimes thirty or forty, as unsaleable or much injured. This man stood as severe an ordeal of cross examination as any witness that has been called. He was asked, who were his masters?—what was the name of the other foreman?—and, what was the name of his colleague who looked over the cloths with him? I expected, and hoped from these questions, that they would all have been sent for; unfortunately they were not, and this was the only reason why I forbore calling the foremen of two other eminent houses in the same line, who were put down as witnesses on my brief. Mr. Lambert, I repeat, stood a most severe cross examination; it was endeavoured to make him say he did not know the gig-dressed from the hand-dressed cloth, except by the place which they came from, though that would have been enough, because the gig-masters are as well known from the hand-masters, as the distinction between these gentlemen sitting here and myself. The gig-cloths are known as coming from a gig-master, and the hand-dressed cloths as coming from a hand-master. He is asked if he can tell from his *judgment*, and he shews you that he can to a moral certainty. How, it is inquired, can he distinguish the one from the other? He says, that the gig-cloth is looser; it feels more hollow in the hand, and the injuries are of a different nature from those which he finds in the hand-dressed cloths. He examines all the hand-dressed cloths which come from an eminent manufacturer in the West of England, and in the course of the last two years he has not had occasion to return five of them; while, on the other hand, he is constantly obliged to object to a great number of the gig-dressed cloths. Well, but do not the drapers take them notwithstanding? Yes, they do, when in want of particular colours, and at particular periods when the demand is too pressing to wait. I do not, says he, absolutely put them on one side unless they are materially damaged, and even then, the drapers will sometimes have them fine-drawn, and put up with them, rather than not have the article for sale.

It is thus proved, not only that damaged goods are bought, but that this argument of self-interest is not a security against such damage, the want of the article is great and pressing, and, whatever may be its state, it is folded up and sent to the draper in London. But can you, as legislators, interested in taking care that the article shall be good in its nature—can you avoid feeling the circumstance of this great proportion of deteriorated cloths out of the number stated?—And, does it not forcibly strike you, that though the urgency of the demand may compel the trade to pass over such defects rather than not have cloths for the execution of their orders--yet, that when peace returns, and when there shall be a choice of markets, the consequence of exporting such cloths may lose us the foreign purchaser?

You have also obtained another extremely important fact from this witness. You have learned from him, that the cloths, whether over-strained at the tenter, or by gig-dressing, go abroad in that state of extension:—those intended for home consumption, especially for the London wear, are damped and cold-pressed, the damping drives them in, perhaps, half a yard upon twenty; but the cloths meant for the foreign market are not damped, but merely pressed, which does not drive them in more than a very few inches at most, so that in whatever state of extension they come from the manufacturer, in that state, or nearly so, they go abroad; the foreign purchaser thus paying for as much more than he ought, as they have been extended beyond the legal length. I ask, can the foreign trade long survive such practices as are here indubitably proved? You remember, that Mr. E. one of the Yorkshire masters, imagining himself not to be within the limits of the stamping act, admits that the coarser kind of woollen cloths are strained in the proportion of five yards upon twenty-five. The article could not be made, he says, without this extension, though he adds, for our comfort, that it goes in again by degrees in proportion as it becomes exposed to the air. But, suppose it is sent away instantly, and shipped before it has time to be thus exposed, it does go in to be sure, and even then it goes in by degrees; but it is as our great coats go in, that is, upon the wearers backs! I had a great coat made the other day, so large, it appeared almost ridiculous, yet the first symptom of a shower which I experienced, a mere misling rain for an hour or two, so shrunk it, that in the evening I protest I could scarcely button it. Mr. E. says, truly, that these cloths shrink back towards the legal length every time they are exposed to the air; but, unfortunately, our character as a great trading nation suffers in the mean time, and diminishes with every yard of cloth that goes abroad.

I come now, Sir, to the third part of my subject, namely, the searchers and sealers. It may, perhaps, be put to me as it was put to one of my witnesses :—Do you imagine that the cloth-workers and sheermen care three farthings about the searchers or the length of their cloth, or the injury done by the sheering-frame. Have they, who are mere workmen, any feeling or care whether the cloth is spoiled or not? why then do they make the searchers and sealers so material a point of their case? My answer is, that these rich master clothiers, who in a few years will be all great men, have less interest in the reputation of the fabric than we who are workmen; we are bound down, as it were, by fate; we have no hope beyond that of obtaining a comfortable return for an honest and laborious life; we have no prospect of employment but according to the pre-eminent reputation of the manufacture; we have nothing to look to for ourselves and our children but the woollen trade, and the being occupied therein; and, therefore, its reputation is a concern of the nearest kind to us, and we conscientiously believe, that if one thing more than another can maintain that reputation, it is the vigilance of searchers and sealers; we therefore urge you rather to give new life and strength to those salutary laws which appointed them, than to listen to the selfish suggestions of those who are compelled to act fairly by the public when these men do their duty. Our ancestors were of opinion, that searchers and sealers were useful; and, therefore, from very early times up to so recent a period as the 5th George the Third, they have thought it wise to appoint and to continue them; the evidence before you has, beyond all possibility of contradiction, proved that these men are highly useful, and it has been shewn that where they attend and are supported in their duty, there has been no illegal and improper straining; but that where they are known to be in the habit of non-attendance; where they are complaisant enough to stay away upon an understanding with the masters, the direct contrary takes place. If you believe the witnesses, you must be satisfied that the searcher being present and doing his duty, when the cloth is taken from the tenter, prevents the iniquity of illegal straining, if so, is not that enough to establish the necessity of continuing those laws which enjoin their attendance?

I questioned one of the masters in this manner:—"If an avaricious master were so disposed, could he not have his cloth strained as many as two yards upon twenty, beyond the legal length, without injuring the appearance of the cloth?" "It would not be to his interest:" was the answer I received. "Never mind interest. Have the goodness to say whether if the master should be so disposed, he could not have his cloth overstrained so many yards!" "I do not know:" was the answer. "But if the master should be so disposed, (I again asked), and the men had their proper cue, could not he have his cloth strained, even when dressed by hand, one or more yards upon twenty beyond the legal length?" At last we got an answer; Mr. D. who went so smoothly over the stones through out the whole of his evidence, and whom I did not choose to interrupt, knowing the admission I should wring from him; even Mr. D. at last, acknowledged, that if a master were so disposed, the thing might be done. I believe

Mr. D. is what is called a crack-master; that he is more particular about his cloths than any other man, and I believe also, that not one of the iniquities I charge are imputable to his mode of conducting his business.

But still this gentleman, you see, when further urged, admits that the thing could be done if the master were so disposed. If then the thing can be done, will you not, as legislators, prevent it? or rather, I ask, will you not refuse to listen to those applications which desire you to take off all restraint, and repeal the prohibitory laws?—which desire you to proclaim it in every foreign market upon the Continent, that our manufacturers of woollen cloth are at length relieved from all legal obligation, and that they must henceforth trade with the British clothier upon the principle of universal suspicion; that they must believe every Englishman to be a rogue and deal with him accordingly; that they must henceforth examine well the length of their cloths, for that the manufacturers are no longer under restriction; that every thing being sure to find its level, self-interest is henceforth to be looked up to as the standard for security! Mr. D. as I have shown, admits the thing may be done—he admits more—he admits that some of the cloths are in fact strained above one yard in twenty more than the legal length. But then, says he, we put substance into our cloths accordingly. Perhaps they may, but that at least proves the illegal straining to take place, which is enough for my argument, it is drawn from themselves, and shows the necessity of continuing legislative prevention.

Another species of iniquity takes place, which I own I was not aware of till lately; nor should I have known it now but for that candour and honour with which Mr. Law Atkinson has conducted himself throughout the whole of the business. It had been intimated to me that in proportion to the understanding between the master and journeymen, this sort of fraud takes place, namely, that when cloths are taken from the tenter, though notoriously overstrained, nevertheless the length is found to correspond with the seal put on at the fulling-mill. I had been informed, that by some management at the fulling mill, it was not unusual to put on seals there denoting a longer length than the cloth actually measured, so as to allow for subsequent straining. Though this had been suggested to me by my clients, I own I could hardly venture to put the question, doubting, myself, the existence of so impudent a fraud; at last I put it fairly to Mr. Law Atkinson, whether it was not so. He admitted his belief that such a practice did take place, and perhaps too frequently amongst fraudulent masters; or else, says he, how could those accounts which I myself put in, shew a less degree of ultimate length after all the process had been gone through, than appears from the seals put on at the fulling-mill? unless something of this kind had been practised upon ourselves, such would have been impossible. You will find this, Sir, one of the greatest sources and items of fraud; and that though it is true that the clothiers are within the letter of the law, when their cloths are not stretched beyond the length sealed at the fulling-mill, yet you will be satisfied from the evidence of that gentleman, and from other evidence which appears on your minutes, that a false seal is frequently put on, which is meant to allow for the illegal straining and stretching of the cloth. Now, will a legislative committee endure that those legal restrictions shall be repealed which can alone prevent such imposition upon the foreign purchaser and the domestic consumer. I must not be told that the difficulty lies in getting men to do their duty; does not the same difficulty lie with respect to the customs and excise? Such an objection is about as weighty as the universal argument of self-interest; which, if trusted to, would keep the manufacturers in no better order than universal philanthropy, or any other principle of mere public spirit. I must not, therefore, I say, be told that you cannot get officers to do their duty. How do you get them to do their duty in the other departments under government, but by introducing such checks, and making such arrangements as to render it almost impossible for fraud to escape detection? It is no argument to say the thing would be very good if you could enforce it, to lament over your imbecility, and to deplore your want of power. Pay the searchers better. The clothiers well know, and it is in evidence before you, that the searchers have not the common means of living, if they do their duty—that they are cast for sustenance upon betraying their trust—that their families must starve if they were faithful, so scanty is their allowance; while, on the other hand, he is liberally rewarded if he refrains from visiting them, or makes his visits in a complaisant way. The allowance which they must trust to for their livelihood is not for looking over the cloths, but for *their non-examination.*

The utility of the searchers was further, though very unwillingly admitted by another of the masters: I think it was Mr. K. I asked him if the searchers visited him sometimes?—he was of opinion that they are a very troublesome impertinent set, and that the sooner they are got rid of the better. In what way, said I, have they been troublesome to you? Why are you so anxious to get rid of these men who so seldom come near you? —why they came once, and one of them threatened to inform against me because my cloth was without the seals, but I set him at defiance, says he, for I knew the cloth had been sold and returned, and it was not necessary that returned cloth should have the seals. I knew the searcher was in the wrong box, and I told him he might do his worst. But suppose, said I, Mr. K. that you had been a man of a different character to what you undoubtedly are, and that a real fraud had been meditated, would not the vigilance of the searcher upon this occasion have prevented it? Yes, said he, if I had meant to be fraudulent, he would certainly have prevented me.—Then, as there are some men to be found in all trades who will be fraudulent, is it nothing, I ask, for the pre-eminence of our staple manufacture, nothing for our reputation abroad, and our consequent export trade, nothing for our consumers at home, to take care by the due maintenance of legislative provisions, that this commodity be produced in the utmost possible degree of perfection, and that fair and honest dealing should be, as much as possible rendered inseparable from our commercial character.

Sir, this brings me to the last division of my case, namely, the law and system of apprenticeship. I do not feel myself under the necessity of going very much at large into it. I stated in my opening address, that the abrogation of the system of apprenticeship appeared to me to be one of the most awful propositions ever submitted to the legislature. It is not enough to shew that a weaver or a cloth-worker can get a knowledge of his trade in a few months or a few years! How long, I ask, is a boy learning to be a haberdasher or a grocer, or any other retail trade? They require no skill, and yet apprenticeship to them is universal; and who shall say that the superior morals of our countrymen have not been owing to the prevalence of a system which places youth under proper controul during that most critical period of life.

when even parental authority is not so effectual as the authority of the master? A parent cannot, consistently with his fond feelings, exercise that strict authority over his son which the master does over his apprentice, from fourteen to twenty-one. During this important period, the master is the fittest governor, and I cannot but impute all the real, substantial, and moral happiness of the commercial part of mankind, to the most excellent system of apprenticeship. It is a custom which has prevailed time out of mind, till within these few years, that in some parts of the country it has been relaxed, principally owing to the introduction of machinery, and the factory system. It will be seen whether even machinery cannot be administered with as much effect, and the child at the same time derive all the advantages of a rational system of apprenticeship. Apprenticeship is not a thing of yesterday, it is like the common law of the land: it has existed from time, to the contrary whereof the memory of man runneth not. I know no period of its commencement. The Act of Elizabeth does not create it, but recognises its long existence, and professes to bring all the laws with respect to apprentices into one Act, the more effectually to enforce them. The Act of Elizabeth may be as obsolete now, as those laws were which it reviewed, when it re-enacted such as were wholesome, and introduced such new ones as experience had rendered necessary; but the system has existed throughout all British antiquity, because it is both a moral and political institution. Will any man say that apprenticeship is confined to commercial views and commercial objects? Will any man say that it ought to depend upon a mere motive of trading, or manufacturing policy? Will not every person who has thought upon the subject, who has read a single treatise upon political economy, or has looked through the acts of our ancestors, admit that one of the great reasons of apprenticeship is, that youth should be under a particular moral restraint during the most dangerous period of their lives, rather than be left at large, as proposed by those who seek the repeal of the apprentice laws? Was it not evidently meant that they should receive the rudiments of their trade in one service, and under one master, in preference to being allowed to work for whom or where they please, as now proposed; and this at a time of life when the best of us are said proverbially to be unfit to be our own masters; at a time of life when even among the higher circles, every pain is taken to interpose authority between youth and its passions, between the temptations with which it is beset, and the liberty to plunge into vice. What rank of society is there so high as not to consider the youth of its class as fit subjects of some restraint and authority different from that of the parent? and is not the fondest parent the most anxious to place them in this state of temporary subjugation? Who then shall have nerve enough to say that by one act of legislative licence they will set afloat the great body of the passions, of the most helpless part of society; that part which, from poverty and the want of education, is the most subject to error, the most liable to temptation, and the least capable of resisting it? What! while law and policy continue to enforce apprenticeship among the better classes of the trading community, as is the case throughout the whole city of London; and while over high born youths you impose guardians, tutors, and curators of all descriptions, will you expose the little helpless progeny of our working population to those ills of soul and body which you tremble to think of when applying the danger to those whose happier fortunes afford them every means of instruction and security? I submit to the, committee that whether looked at in a moral or political point of view, the strongest objections present themselves against that freedom from restraint which you are urged to enact. If you give way in this instance, it would be absurd to argue that the abrogation of apprenticeship will be confined to the woollen trade, and that it will not extend to all the various other branches of manufacture. No man can imagine for a moment that you can do away apprenticeship among the clothiers, and support it among the manufacturers of leather, iron, and other articles. You must annihilate the system altogether, or you must have the courage to enforce it in this manufacture as well as in others. I do not mean that you need enforce it according to the extent of the statute of Elizabeth, that is in many respects an obsolete statute. I do not say that it is absolutely necessary that apprenticeship shall be for seven years. I do not say that you shall not listen to the observation of an honourable member, which I felt forcibly when it was urged, namely, that when one youth is bound an apprentice, and sees another working by his side who is not bound, getting five or ten shillings a week, while he is getting nothing, he will feel great uneasiness and impatience. In order to legislate wisely, you must undoubtedly legislate according to the temper of human nature. But who will deny that a system of apprenticeship might not be adopted, improved by all that experience can now suggest. If seven years be inconvenient, say six, or even five years. If it be necessary to give to youth progressive wages, in God's name let it be so; but surely all this might be done consistently with that degree of domestic relationship and moral restraint which I have stated as absolutely essential to the temporal and eternal welfare of our British youth!

I have said I would rather read the sentiments of others, than trouble you much with my own. I will state in a few words what Mr. Justice Blackstone says upon the subject, it is in vol. i. of the Commentaries, page 428.

"Persons serving seven years as apprentices, to any trade, have an exclusive right to exercise that trade in any part of England. This law, with regard to the exclusive part of it, has by turns been looked upon as a hard law, or as a beneficial one, according to the prevailing humour of the times, which has occasioned a great variety of resolutions in the courts of law concerning it, and attempts have been frequently made for its repeal, though hitherto without success.

"At common law, every man might use what trade he pleased, but this restrains that liberty to such as have served an apprenticeship The adversaries to which provision say, that all restrictions (which tend to introduce monopolies) are pernicious to trade. The advocates for it allege, that unskilfulness in trade is equally detrimental to the public as monopolies. This reason, indeed, only extends to such trades in exercise whereof skill is required; but another of their arguments goes much further, namely, that apprenticeships are useful to the commonwealth, by employing of youth, and learning them to be early industrious; but that no one would be induced to undergo a seven years apprenticeship, if others, though equally skilful, were allowed the same advantages, without having undergone the same discipline: in this there seems to be much reason."

Ours is not the only country that venerates this system. We are continually quoting the wisdom and the enterprise of the French. The law of apprenticeship was infinitely more rigorously administered there than here. I find the following passage in Postlethwaytes translation of Monsieur Savary's Commercial Dictionary, who evidently accompanies the text with sentiments of his own, viz.

"The commerce of a nation depends on the excellency of its manufactures, and those again on the skill, ingenuity, and dexterity of its manufacturers, and artificers of every kind. On these considerations we may presume that the French have made such laws to oblige their artists to make themselves thorough masters of their respective trades, and to exclude them from the freedom of the several corporations of arts and trades, unless they are capable of performing with their own hands what they call a master-piece of workmanship, that is such a piece of workmanship as will shew a young man master of his business, and this must not be done in a corner, but in the presence of those sworn for the purpose.

"The national benefit of this policy will further appear, from considering the disadvantage Great Britain must inevitably sustain in her commerce, were her manufactures to degenerate in their ingenuity and dexterity.

"However slight some people may make of these considerations unon apprentices, yet the breeding up our young people properly to trade, in every branch of it, seems a matter of the last consequence to a trading nation, and may one time or other be found worthy some people thinking of, when they are once convinced of this important truth, that traders alone are the source of all their treasures, and consequently of all their splendor and magnificence.

" Since such exquisite skill in the manual operation is required by all any way practically concerned in the woollen manufacture; do not all our national advantages arise from the woollen manufacturers, who are to act in every part hereof, for they are not bred in *a few years*. Ought not the wisdom of the nation to be alarmed at the daily artifices which are used to decoy and instigate our artists and manufacturers out of the kingdom, to the emolument of other countries, and the certain ruin of this? And can any thing effectually do this but giving them all due and reasonable encouragement?"

It was apprehended by an honourable Baronet to whom it is impossible not to pay the greatest attention, that when we talk of maintaining apprenticeships, we, in fact, mean to prohibit the infautine classes from working, as they do now in factories;—Not at all; they may continue to be employed till they arrive at an age fit to be put apprentice to learn the trade, which they will do so much the sooner from their previous experience. But what I contend for is, the continuation of that system of apprenticeship which binds the youth from fourteen years of age to twenty-one, unless you should think proper to qualify the term of its duration. It appeared, by the evidence of 1803, that in the West of England there is scarce a cloth-worker or sheerman who has not served seven years to the trade. The system, it is true, has been much relaxed in Yorkshire. In Gloucestershire there is a substitute for it, which they call colting. I put this question to several of the masters :—" If a system of apprenticeship were adopted, giving a youth progressive wages according to his skill and merit, but still keeping him under the eye and authority of his master, would it not be infinitely better, and more conducive to his moral and general welfare, than allowing him, as proposed, to work for whom he pleases and for what he can get ?"

The answer was always in the affirmative. But it was added, that masters would not be troubled with apprentices. I admit that this objection may apply to the higher classes of modern masters, and I regret that it does, but it does not apply to the general description of masters in the woollen trade, the mass of whom are not so elevated or so fashionably averse to domestic trouble and duty. Allow me to observe, that you are sitting here to inquire what may be best to legislate for the future, and if you should think that some improved plan of apprenticeship would be infinitely preferable to the annihilation of that almost paternal system, you will recommend it with such other improvements as may suggest themselves for the future government of the woollen manufacture.

But, take it, that persons in elevated stations may say, We cannot endure these troublesome rogues under our roof. Let them, in that case, discharge their duty by deputy—let the foremen, or others whom the masters may appoint, receive the apprentices into their houses and families, and exercise over them. as far as may be, a delegated authority. Is there not ten thousand degrees of difference between that and letting these little premature journeymen and journeywomen roam at large, and be at their own disposal? It seems to me that if the foreman, as is not unusually the case, or other honest and worthy people were to receive three, four, or half-a-dozen at a time, of these young people, into their respective families, it might be rendered an object to themselves, and would get rid of one of the principal objections made by these sensitive masters to the burthensome part of apprenticeship.

I beg it to be recollected, that not one witness has ventured to say there could be any material difference in the term now considered as necessary for the bringing up of a *cloth-worker and sheerman.* Even Mr. Law Atkinson speaks of three or four years as the shortest time in which a lad could be fitted for that branch of the business; so that the question, as it relates to my clients, remains almost untouched, respecting the necessity of an apprenticeship for seven years. It does not at all follow, that masters should not employ children as they do now, till they get to be thirteen or fourteen years of age, but when they arrive at that age let them become domesticated—let them then, at least, be under moral government, and learn the na ture of the manufacture in its various branches as their fathers have done before them, through the medium of a legal apprenticeship.

These are the points which I have thought it my duty, as generally as I could, to submit to the committee.

When in the first and second place I took the liberty to call your attention to the subject of machinery, I did not think it necessary again to notice the cruel and malignant falsehood which I have so often refuted, namely, that my clients oppose machinery *in general*, there is not a member of this honourable committee but now fully understands; and I hope, that from the fidelity of their report it will soon be understood by the House and by the whole public, that the only machines which we object to the toleration of, are those, which in letter and spirit, stand already prohibited by law. I implore the committee to recollect that machinery may be abused as well as used. I beg they will have the goodness to consult those authorities which have drawn the line between the one and the other; and, above all, to recollect not only that *no necessity* in a political point of view has been shown for the gig-mill and shearing-frame, but that they have been proved to demonstration to be in themselves absolutely as injurious to the fabric now, as they were in the reigns of Edward the Sixth, and of Philip and Mary, when the former was specifically put down, and the effect of the latter deprecated and prohibited. Nothing can be more fully proved than the fact of their depopulating tendency, except those other facts, which we have equally proved, namely, that they yield no adequate compensation for this great public evil and private calamity, either from the quality or quantity of the cloth which they dress.

I endeavoured to show, in the third place, the advantage of continuing, invigorating, and supporting a system of searching and sealing, and I have most abundantly proved, that where the officers have attended to their duty and visited the tenters, the cloths have not been unduly stretched; while, on the contrary, in those instances in which they have been seduced, or terrified from asserting their function, or, as has been the case for the last three or four years, been persuaded the Suspension Bill forbids their interference, such a deterioration of the commodity has taken place, as must, if the present relaxation from the searching laws were to continue, deeply affect your foreign trade, and that perhaps at no very emote period. If the mere relaxation of those laws have done so much mischief, what have we not to dread from their proclaimed repeal? from that complete release from all restraint, now so earnestly called for by the master-clothiers and their philosophical advocates? Sir, in the fixed opinion that the woollen manufacture, considered as the staple of England, cannot long survive the proposed alteration of system, I confidently assure myself that it will not have the countenance of this honourable committee.

I lastly submitted, That, profiting by experience, that best of teachers, a system of apprenticeship might be introduced, free from the objections which have been urged, and yet calculated to embrace all those advantages of a moral, social and religious nature, which I so earnestly alluded to, and which I contend were so anxiously provided for by our ancestors, and the wisdom of which the greater part of the kingdom, even up to this day, recognizes by its practice. If, upon this latter subject, I have gone beyond the cold rules of argument, impute it, I beseech you, to feelings which I could not restrain, I have but obeyed the impulses of the man and the Christian, in lifting up my voice against the abrogation of a system to which the moral world owes so much. Regarding its abrogation, or which is the same thing, the repeal of its privileges, as a dissolution of those bands which have hitherto held the trading part of this trading country together, as breaking down those barriers which have been interposed to check and prevent early licentiousness, the sure result of early emancipation, and as tending to effect a revolution of the worst kind, in the character of a vast proportion of my countrymen; I will continue to protest against such a change before whatever tribunal I may be called to argue the question, I will at no moment forget that I am a citizen as well as an advocate, or fail to remind my judges that the proposed repeal of the laws of apprenticeship, while it sinks the immense body of men, for whom I appear, in utter despair, cannot confine its pernicious effects to the woollen manufacture, but must pervade the whole kingdom, and successively operate upon every other branch of trade.

You are seated here as judges upon this question—whatever may be the report you agree to, it cannot but have great, and, I trust, deserved weight—Should you in that report suggest some medium measures, some such qualifications as may meet the interests, sooth the feelings, conciliate the affections, and even, as far as it is safe so to do, indulge the prejudices of the respective parties before you, it will have the best and happiest effect. But if, in the spirit of stern authority, a bill should be introduced expressly to gratify one of the parties, I believe it will have the most fatal tendency, and as such I solemnly deprecate it. I do not mean to intimate that my clients will want temper, their conduct has proved the contrary; it is now three or four years that they have been under what I may truly call my paternal guidance and direction, and I can boldly and conscientiously assert, that their conduct throughout has been most exemplary; I have personally the means of knowing, that from the introduction of the first bill of repeal, which happily was thrown out, they have taken no public step whatever but with the knowledge and advice of his Majesty's ministers, or of those noble and eminent persons who so graciously interposed on their behalf, who admitted them to audiences with the kindest condescension; who received them into their closets while higher born men waited in the anti-chamber; who lent the most patient ear to their representations, and reasoned with them upon their case with parental mildness. This conduct had the effect it ought to have, it has left an indelible impression of gratitude upon their minds, and I be-believe, there is not one of the noble and honourable persons I allude to, for whom these men would not lay down their lives. Sensibility is not confined to station, as was made apparent this very morning, by the effect of the resolution to which you condescended to come, it filled the petitioners with the highest respect for you, and in a moment banished, I hope for ever, all remembrance of the asperity with which they imagined themselves to have been treated in the early sittings of this committee. Be assured, Sir, that his Majesty has in no part of his dominions a hundred thousand of more loyal subjects than those concerned in the woollen manufacture; it remains with yourself and your honourable colleagues to render them the happiest, and make them the best!

I am aware, Sir, that from an extreme anxiety to compress my address within the time allotted to this days sitting, I may have omitted many important points upon so extensive a subject, and that the same cause may have rendered my arrangement much less perspicuous than I could have wished it to have been. It is my consolation that I address those whose long and laborious inquiries have made them masters of the subject, whose judgment therefore will supply my defects, and whose urbanity I know will excuse them. To this request, allow me, Sir, to add my grateful thanks to the committee for the attention with which they have been pleased to honour me.

THE END.

C. Stower, Printer, Paternoster-Row.

British Labour Struggles: Contemporary Pamphlets 1727-1850

An Arno Press/New York Times Collection

Labour Problems Before the Industrial Revolution. 1727-1745.

Labour Disputes in the Early Days of the Industrial Revolution. 1758-1780.

The Spread of Machinery. 1793-1806.

The Luddites. 1812-1839.

The Spitalfields Acts. 1818-1828.

Friendly Societies. 1798-1839.

Trade Unions Under the Combination Acts. 1799-1823.

Repeal of the Combination Acts. 1825.

Trade Unions in the Early 1830s. 1831-1837.

[Tufnell, Edward Carlton]
Character, Object and Effects of Trades' Unions; With Some Remarks on the Law Concerning Them. 1834.

Rebirth of the Trade Union Movement. 1838-1847.

Labour Disputes in the Mines. 1831-1844.

The Framework Knitters and Handloom Weavers; Their Attempts to Keep Up Wages. 1820-1845.

Robert Owen at New Lanark. 1824-1838.

Motherwell and Orbiston: The First Owenite Attempts at Cooperative Communities. 1822-1825.

Owenism and the Working Class. 1821-1834.

Cooperation and the Working Class: Theoretical Contributions. 1827-1834.

The Rational System. 1837-1841.

Cooperative Communities: Plans and Descriptions. 1825-1847.

The Factory Act of 1819. 1818-1819.

The Ten Hours Movement in 1831 and 1832. 1831-1832.

The Factory Act of 1833. 1833-1834.

Richard Oastler: King of Factory Children. 1835-1861.

The Battle for the Ten Hours Day Continues. 1837-1843.

The Factory Education Bill of 1843. 1843.

Prelude to Victory of the Ten Hours Movement. 1844.

Sunday Work. 1794-1856.

Demands for Early Closing Hours. 1843.

Conditions of Work and Living: The Reawakening of the English Conscience. 1838-1844.

Improving the Lot of the Chimney Sweeps. 1785-1840.

The Rising of the Agricultural Labourers. 1830-1831.

The Aftermath of the "Lost Labourers' Revolt". 1830-1831.